Back-to-Basics Audio

Julian Nathan

Newnes
An imprint of Butterworth-Heinemann

Boston   Oxford   Johannesburg   Melbourne   New Delhi   Singapore

Newnes is an imprint of Butterworth-Heinemann.

Copyright © 1998 by Butterworth-Heinemann   076384233
℞ A member of the Reed Elsevier Group

GLOBAL RELEAF 2000 Butterworth-Heinemann supports the efforts of American Forests and the Global ReLeaf program in its campaign for the betterment of trees, forests, and our environment.

ISBN 0-7506-9967-1

British Library Cataloguing-in-Publication Data
A catalogue record for this book is available from the British Library.

The publisher offers special discounts on bulk orders of this book.
For information, please contact:
Manager of Special Sales
Butterworth-Heinemann
225 Wildwood Avenue
Woburn, MA 01801-2041
Tel: 781-904-2500
Fax: 781-904-2620

For more information on all Butterworth-Heinemann publications available, contact our World Wide Web home page at: http://www.bh.com

10 9 8 7 6 5 4 3 2 1

Printed in the United States of America

Contents

Introduction

This is the book I wanted when I first started work in audio. It is written for hands-on technicians, and also for other interested people who just want to acquire the basics of audio and electrical principles without completing a college course. It's light on formulae and calculation, but intends to bridge the gap between practice and theory, and includes a range of relevant facts and events. The few essential formulae presented can be worked with a pocket calculator, yet they are the foundation of all electrical, electronic, and acoustic work.

The subject is audio, and how systems are put together and calibrated. It does not cover the manufacture and service of system components, so specific design and repair detail of source equipment, processors, amplifiers, and speakers has been omitted. However, the attributes of amplifiers, speakers, test equipment, and tools are discussed to assist the process of choosing them to suit different applications. A basic idea of how things work is also given to enable the reader to make loss pads, passive equalizer networks, power supplies and relay systems, because they are the interface and support units for the major parts of a sound system.

Wherever the opportunity arises, a parallel description of electric power and allied technology is included, because it helps to illustrate that the fundamental principles of audio and power are exactly the same. The chapters on drawing and construction show how to design racks and panels, understand complex systems, prepare working diagrams, and communicate ideas so that others will immediately recognize the function of the systems you design.

Aspects of movie soundtrack recording, signal processing and acoustics are discussed in detail, subjects that relate in a practical way to to the reader's appreciation of professional and domestic audio systems. Wherever I've worked in audio, people have asked for explanations. Here are some of the answers.

Julian Nathan joined the audio service and manufacturing industry in 1954 and moved into motion picture engineering and production in 1960. He installed and operated recording theaters in Sydney, Australia, and set up similar facilities in Papua New Guinea and Hong Kong. Later he spent several years in cinema installation and service including audio visual theaters and government sound systems, then worked in the hi-fi and professional sound equipment markets. His particular interest is home theater.

Back-to-Basics Audio

Chapter 1.

Electrical Principles

AUDIO'S THREE PARTS

Audio is Acoustic, Mechanical, and Electric. Sound, or wave-motion in air, is sensed, processed, and reproduced by electro-mechanical and electronic devices. It wasn't always electric, but that's the medium we use to do almost everything, and do it better.

Sound is **transient**, it's gone as soon as it happens, but it can be examined and its equipment tested, largely with continuous tones, or in the **steady state**. More details of the nature of sound are given in Chapter 4, but as a necessary background to the technology of audio, the briefest account of electricity and its associated parts occupies most of this chapter.

ELECTRONICS AND ELECTRICITY

All matter is made up of identical Neutrons, Protons, and Electrons. In various combinations they form atoms; the basic clusters that form the elements. In turn, one or more atoms make up a molecule, the smallest functional form of an element or compound. The atomic weight of a stable element is determined by the number of negative charge electrons circling the nucleus of its atom, which in turn is made of an equivalent number of positive charge protons, plus various numbers of neutrons.

Electrical conductors have loosely bonded electrons in their atoms. Electrons are able to flow from atom to atom in conductors, and since they have a negative charge, convention says that positive current flows in the opposite direction to electron flow. Electrons flow from negative to positive in conductive media because unlike charges attract, so current flows from positive to negative. Poor conductors have closely bonded electrons, making flow difficult. They have high **resistance**.

All the use we make of **electronics** is based on working with the outer electron shell of atoms. We can make electrons pass from atom to atom. By heating conductors we can make them incandescent; photon emission is an indication of super-excited electrons. In fluorescent, carbon and xenon arc, and neon lamps, we see a demonstration of **plasma**; conducting, ionized gas; the fourth state of matter. **Maser** and **laser** technology, including semi-laser devices like LEDs (light emitting diodes), proves the connection between excitation of electrons by one form of energy, and production of microwaves or coherent light (single frequency or narrow band photon emission) by the electrons thus stimulated.

Nucleonics, on the other hand, is concerned with mobilizing the

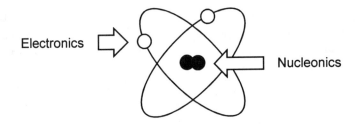

neutrons and protons that form atomic nuclei, in order to release energy. Nuclear engineering is beyond the reach of most people because it's expensive and difficult to handle.

Electronics, understood and thoroughly exploited, is easy to get into, and safe as a box of matches.

AN ELECTRICAL CIRCUIT

1. When the switch in the circuit of Fig. 1.1 is open, *voltage* is present *across* the battery and therefore *across* the switch. But no current flows.
2. When the switch is closed, *current* flows through the (lamp) load, limited by the combined *resistance* of the load, the source, the conductors, and the switch.
3. Work (Watts) done in the load equals electrical pressure (Volts) times current (Amps). A small proportion is wasted in heating the other resistive components, including the battery.

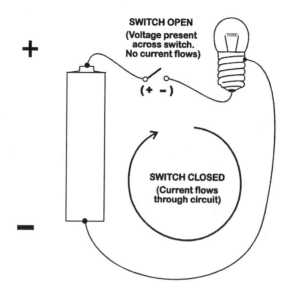

Fig. 1.1 **An electric circuit.**

4. The current (flow) at any point in the circuit is the same at any other point (in a simple circuit that does not branch).
5. Electricity requires a complete circuit for current to flow.

ELECTRICAL PRINCIPLES

Volts, Amps (amperes) and **Ohms** are inter-related quantities, and refer to the physical properties of Electricity. Without an easily acquired understanding of these three basics, very little progress can be made in any application. A working knowledge is given in the following few pages, containing nothing more complex than simple equations.

Voltage is **Electrical Pressure**. Just as water in a tank exerts physical pressure on the pipe it supplies, so a voltage source exerts electrical pressure, or a potential difference, at the parts of a circuit to which it is connected. Voltage is also referred to as **Electromotive Force**, or EMF.

VOLT : The unit of ELECTRICAL PRESSURE

Symbol : V

Amperes denote **Electric Current Flow**. When pressure is released, water in a pipe flows. The flow rate could be expressed in liters per minute. A fundamental observation is that the flow rate is the same at any point in the pipe, assuming it does not branch. Three liters per minute at one end guarantees three liters per minute at the other. Electric current flow is the same.

AMP : The unit of ELECTRIC CURRENT FLOW
Symbol : I

Ohms are units of **Electrical Resistance** (to current flow). Again consider the water pipe. If it is rough inside, or narrow, it could be said to have high resistance. Even if it is perfectly smooth, it will have *some* resistance. Resistance is also a feature of Electrical Conductors.

The degree of resistance depends on the material and cross section area of the conductor, at normal temperatures.

OHM : The unit of ELECTRICAL RESISTANCE
Symbol : R

OHM'S LAW

Current flow is the operative factor in an electric circuit. Voltage makes it happen. Resistance retards it. This all works by good fortune, since without resistance, the simple act of switching on a light would initiate enormous currents, and other means would have to be found to limit the flow. Resistance is like friction, without which we would all need spiked shoes to stand erect.

Volts and amps are related by the equation: $I = \dfrac{V}{R}$

AMPS EQUALS VOLTS DIVIDED BY OHMS

This is the basis of Ohm's Law.

Current flow is directly related to Voltage, but inversely to Resistance. More Volts, more Amps, but more Ohms, less Amps.

USING OHM'S LAW

The equation can be converted to determine any one quantity, given the other two, by the rules of simple equations:

A figure can cross the equals sign
provided it also crosses the divide line.

Therefore: $R = \dfrac{V}{I}$

OHMS EQUALS VOLTS DIVIDED BY AMPS

By the same equation, $V = I\,R$

VOLTS EQUALS AMPS x OHMS

This classic formula is valid under all conditions. It applies to materials that change resistance when they get hot, and equally to AC circuits where **R** represents **impedance**, a combination of resistance with inductive and capacitive reactance (see page 7).

Draw a button inscribed with the following:

Place a finger on the quantity you need to know. Calculate the other two.

VOLTAGE DROP

Referring to the next chapter, on pages 19 and 20, it will be seen that **voltage drop** is the quantity being measured in the majority of cases when a series resistor is the focus of attention. Calculation of a series resistor, for the purpose of reducing the voltage across the following circuit, requires an appreciation of voltage drop, rather than the voltage that results from the resistor's presence, or the supply voltage.

Practical examples of voltage drop calculation appear in Appendix A at the end of the book.

POWER

Volts and Amps by themselves are not *Power*, the part that does the work, but in combination, they will be *Power*.

A WATT IS A JOULE PER SECOND

We won't be much concerned with Joules, except to say that the Joule is one of the basic units denoting **an amount of energy**. Example: a 450 joule photographic flash; 450 joules of energy expended in a millisecond. An amount of energy converted 95 percent into light; 5 percent into heat. Energy, like matter, can neither be created nor lost. If it doesn't light up or get hot, it *must* rotate, make noises, or *do* something.

1 JOULE PER SECOND = 1 WATT

746 WATTS = 1 HORSEPOWER

WATT: The unit that represents RATE OF ENERGY CONSUMPTION (OR CONVERSION)

Symbol: W

A 60 watt lamp consumes 60 watts; 60 Joules per second; 3,600 Joules per minute.

$W = VI$ *WATTS EQUALS VOLTS TIMES AMPS*

Correlating this with $I = \dfrac{V}{R}$, from page 4, we have: $W = \dfrac{V}{R} \times V$ which equals: $\dfrac{V^2}{R}$

$W = \dfrac{V^2}{R}$ *WATTS EQUALS VOLTS SQUARED DIVIDED BY OHMS*

INTRODUCTION TO MEASUREMENTS

Amplifier output power measurement, whether in detail or in quick tests, is among the most often used procedures in audio. The above equation, relating watts to the square of the voltage, will come up frequently in connection with amplifiers and speakers. Given the speaker (or amplifier load) impedance, a pocket calculator will quickly tell the user how many watts will be dissipated in the speaker and its circuit. A resistor is used in place of the speaker to avoid inductive and back-EMF effects when amplifiers are tested, for speakers are generators as well as being motors.

The behavior of different conductor materials is useful in a practical sense. Most increase resistance as they get hot. Other conductors and devices have a negative temperature coefficient and their resistance falls as temperature rises; silicon diodes increase conductivity, the inverse of resistance, as the current through them is increased. Nominal and instantaneous resistance are not always the same. But despite all these inconsistencies, all of which can be put to good use, and need to be considered in measurements, Ohm's Law is still valid at any given instant.

INDUCTANCE, CAPACITANCE, AND IMPEDANCE

Inductance is demonstrated by a conductor's apparent inertia to changes in current flow. In the case of alternating currents, the effect increases with frequency, which is denoted in Hertz (Hz), or Cycles per Second. It is caused by the rise and collapse of the magnetic field which surrounds all active conductors. The field requires energy expenditure to establish it when current flow commences. The same amount of energy is given back when the field collapses; a feature which can be compared to inertia and momentum. The effect can be concentrated by coiling the conductor, and intensified by adding core material. **Chokes, coils,** and **transformer windings** are examples of inductors.

Just as an inductor's magnetic field is established when current is initiated, so too it collapses when the current is interrupted, causing a voltage to appear across it called the **Back-EMF.** Since it often has no load to work into because the circuit has been opened by a switch, the back-EMF can be much higher than the voltage that originally charged the inductor. Inductors therefore can be charged and will give back a charge.

INDUCTORS PASS D.C. BUT OFFER IMPEDANCE TO AC

Unit: Henrys, millihenrys, microhenrys; H, mH, μH

Capacitors are devices with two areas of conductor separated by an insulator called the **Dielectric.** Capacitance is the voltage storage effect which operates when a capacitor is charged by a momentary current. The area and spacing of the capacitor "plates," and the type of dielectric material, determine the capacity and voltage tolerance of the device. Capacitors will charge and discharge rapidly unless slowed by the presence of inductance or resistance in the circuit. This enables alternating currents to pass capacitors, which block the passage of direct current.

Many parts of a circuit have **Capacity,** intentional or otherwise, including cables, due to their proximity, or closeness. A capacitor passes AC more readily as the frequency rises; for example, shunt capacity is a source of high frequency loss in long audio cables.

When a capacitor is charged, a momentary current flows until the capacitor is stressed to the extent of the charging voltage. If the voltage source is then removed, the capacitor will remain charged until the charge either leaks away through internal leakage paths, or is presented to a load, in which case the original charge energy will flow in the opposite direction, discharging the capacitor.

CAPACITORS PASS AC BUT BLOCK DC

Unit: Microfarads; μF or MFD

Audio sources and inputs including speakers have resistance and obey Ohm's Law, but because alternating and signal currents encounter inductive and capacitive reactance as well as ordinary resistance, their circuits are said to have **impedance**, rather than resistance.

Impedance varies with frequency. The Resistance component is frequency independent, but the following attributes of Inductance and Capacitance modify a circuit's behavior at different frequencies as follows:

AN INDUCTOR PASSES LOW FREQUENCIES MORE READILY THAN HIGH FREQUENCIES

A CAPACITOR PASSES HIGH FREQUENCIES MORE READILY THAN LOW FREQUENCIES

RESONANCE

Inductive and capacitive reactance are inverse effects. If their values in ohms are the same at a particular frequency, a circuit is said to be at resonance, or in power mains parlance, it has **unity power factor**. At resonance, or at unity power factor, voltage and current waveforms are in step, or **in phase**, a condition that is charted and described in more detail in Chapter 3. In areas like transformer and filter design, care is taken to control the effect of stray capacities which may produce unwanted resonances in or near the audio band. Many inductive power devices like fluorescent light ballast chokes and electric motors employ power factor correcting capacitors in their circuits to conserve energy which would otherwise be wasted as heat.

A desirable parameter of audio design is that a system should be unconditionally stable up to ten times the pass-band. This means that resonances should be under control up to at least 200 kHz; amplifiers can suffer inefficiency or damage due to oscillation at some frequency beyond audibility. In many audio devices, a non-resonant condition can be acquired by adding resistance to the circuit, which "damps" oscillation, but instability is primarily avoided by employing circuit procedures that avoid phase shifts of more than 90 degrees that may cause positive feedback at some frequency within the response of the equipment.

LINE IMPEDANCE

In any circuit, both series and shunt **reactance-plus-resistance** components make up the circuit's **Impedance**, which is its effective parallel and series resistance to alternating current flow (Fig. 1.2). Unless means are in place to prevent it, circuit impedance is frequency dependent to a degree that may or may not be relevant. For example, using a low impedance source or termination for an audio line will make irrelevant the small capacity occasioned by proximity and length of the conductors. Long lines used for high frequency transmission use other means to make them frequency independent, as will be seen in Chapter 3.

Inductive and Resistive Loads

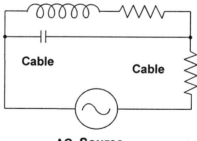

AC Source

Fig. 1.2 The type of termination which may be met by an AC source; for example, a line driver.

IMPEDANCE IS COMBINED INDUCTIVE AND CAPACITIVE REACTANCE AT A GIVEN FREQUENCY, PLUS RESISTANCE.

Symbol: Z

MAGNETISM AND SPEAKER DRIVERS

The magnetic field that surrounds every current carrying conductor has other attributes that make electric motors and speakers possible.

Magnetic poles can only exist in pairs, or as a complete magnetic circuit. Just as two permanent magnets attract and repel each other, so all magnetic fields behave in the same way; like poles repelling, and unlike poles attracting (Fig. 1.3).

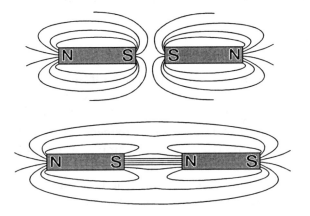

Fig. 1.3 **Magnets in repel and attract modes.**

Because the magnetic field of a conductor is a force that acts **perpendicular** to the direction of current flow, single current conductors, and concentrated conductors like coils, move *across* another magnetic field when free to do so. In a speaker, the voice coil conductors are perpendicular to the field across the annular (ring) gap, so the direction of motion is determined by the audio frequency field of the voice coil conductors relative to the stationary magnet field (Fig. 1.4).

Three factors influence the fidelity with which the voice coil follows the audio signal, and are primary ingredients in the design of a speaker driver, introducing response and phase errors if they are not all effective:

1. The non-varying "excitation" field across the coil needs to be strong enough to match the mass of the cone and the air load.
2. The amplifier output and speaker circuit should have low resistance, so that the voice coil is compelled to closely follow the signal voltage level changes and polarity reversals. Amplifier damping of the

speaker can be compared to engine braking in a car; holding the cone and voice coil when they should be still. Damping factor strongly influences reproduction quality.

3. The cone must offer low inertia to the driving force so that it will accelerate and decelerate on command.

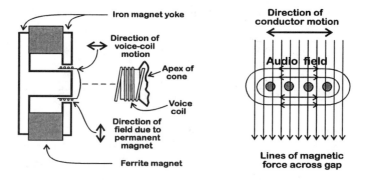

Fig. 1.4 Relationship of voice coil to speaker magnet.

SPEAKERS AND ELECTRIC MOTORS

The functioning of all electric motors, including magnetic speakers, depends on the principle of current carrying conductors moving perpendicular to a **magnetic field**.

Similarly, electric generators work because current is produced in a conductor passing *across* a magnetic field. Speakers also work as generators while they are moving, producing a voltage and current that opposes changes in applied voltage and polarity, hence the importance of amplifier damping.

That is to say, a suitably low amplifier output impedance, coupled to the speaker via low resistance cables, exercises a strong hold over the speaker's movement, not only accelerating it, but also decelerating the cone and its air load as the signal waveform directs.

Because speakers are also generators while they are being driven by an amplifier, connecting dissimilar speakers in series will produce poor results, even if they are the same nominal impedance and size. This is because they tend to drive each other at different frequencies. The problem does not occur with identical speakers because their mass, compliance, and other parameters are the same.

Familiarity with motors can tell us a lot about speakers, because in principle, they are the same device. Motional feedback bass speakers have much in common with servo-motors, for example, whose purpose is to accelerate, brake, and rapidly chance direction on signal command.

They both employ motion and acceleration sensors and negative feedback, and are designed for low inertia. Even a slow moving system like a camera lens servo-motor requires a frequency response up to 300 Hz to ensure stability and avoid overshoot that causes hunting, the condition where the moving mass oscillates around the true position.

THE INVERSE SQUARE LAW AND SPEAKER DESIGN

Non magnetic speakers like piezo-ceramic horn drivers also have a valuable place in sound reproduction, but the inverse square law, which basically states that it is four times harder to achieve operation at twice or half a device's frequency of optimum efficiency, makes it impractical to construct piezo speakers for use in the lower half of the spectrum.

Each part of the audio spectrum calls for its ideal type of speaker driver. In full range single cone speakers, this is satisfied by different parts of the cone moving independently by design, or "cone breakup." The center of a paper fiber cone radiates the high frequencies, and the whole cone diameter, which is loaded at the outer edge by a coupling and seal of suitable flexibility, radiates the low frequencies.

Many design features have been used to augment this effect; the use of specially felted cone material, concentric flexible corrugations pressed into the cone, damped suspension rings, and high frequency radiator cones and domes.

Rigid plastic or aluminum laminate cones have been used for low frequency drivers, an indication that a large cone that does not exhibit "breakup" will not move at higher frequencies because it has too great a mass.

So every technique has its place, due to the immutable existence of the inverse square law, which will not allow us to do everything the same way unless design factors are multiplied many more times than is economically feasible.

The object of much evolution in audio has been to bring a desired effect into range of a frequency band using radically smaller devices, and this means finding entirely new ways to do things, a continuing process that becomes apparent if one examines audio and electronics trends over the last three decades.

BI-POLARITY

It is the nature of physical forces to have two parts: action causes reaction, temperature and pressure differentials compare opposing limits.

Electrical and related quantities are no different: Positive needs negative; Magnetic North poles do not exist without magnetic south poles.

It has been said that static electric charges deny this rule, that a positive or a negative charge can exist by itself. Well, it's certainly possible to isolate a charge in an insulated body, but the question is: Positive or negative relative to what? Usually to Earth or the nearest large body. Storm clouds and capacitors work that way.

Charges relate to each other: 'A' is positive with respect to 'B.' A charged capacitor has two ends; one positive, the other negative. And if there is nothing else to relate to, a charged body can be so many volts positive or negative with respect to ground. The Earth is one half of a storage capacitor, ready to use in conjunction with any other convenient body, like a cloud of water vapor, as in lightning and thunderstorms.

The basic quantities all exist relative to something. Voltage is expressed as a potential difference. Just so many volts has no meaning, but twenty volts AC or DC at this point in a circuit relative to that point describes the situation completely. If current is flowing, **voltage drop** is expressed as the **potential difference** between two points in the circuit.

Bi-polarity is an essential part of Electricity and Magnetism, and ease of using Audio Technology depends very largely on understanding this duality.

DIRECT AND ALTERNATING CURRENT

Direct current flows in one direction, the opposite direction to electron flow, because electrons are a negative quantity. Alternating current flows alternately in one direction, then the other, at a rate determined by its frequency, in cycles per second, better known as Hertz. DC and AC have different capabilities, or attributes. For example, when a circuit is interrupted, direct current tends to continue, and if the voltage is high enough it will form an arc, or plasma conduit, to bridge the gap. Plasma is ionized, conductive gas. AC does not do this so readily because the voltage and current fall to zero at each polarity reversal, and the arc will go out. AC carbon arc lamps work by maintaining a hot ball of ionized conductive carbon dioxide plasma which re-ignites the arc every half cycle. It may be noticed that micro-break switches have different ratings for AC and DC. This is because DC will form and maintain an arc over a small contact separation above a certain voltage, whereas AC requires a much higher voltage to do the same thing, having to actually vaporize contact material to provide ignition for the next cycle.

AC will pass continuously through a transformer or capacitor. DC will not pass a transformer, except when the current starts or stops, because it is the rise and collapse of the transformer core's magnetic flux that induces the corresponding amount of energy in the secondary winding as voltage and current. So pulsed DC (Fig. 1.5) can pass through transformers to a certain extent, assisted by specialized core and winding design. Transformers are described in Chapter 11.

The relationship between RMS and peak voltage and current (Fig. 1.5) is discussed in Chapter 3, in relation to amplifier output measurements. Three-phase electricity supplies (Fig. 1.5c) are mentioned briefly in Chapter 11, partly to stress that AC in power mains or transformers is the same type of energy as speaker and audio line signals.

GRAPHICAL REPRESENTATION OF WAVEFORMS

Voltage and current waveforms, that is, the relationship of voltage or current levels to the passage of time, may remain abstract factors in the imagination until a picture is drawn.

The graphs of Fig. 1.5 show the rise and fall of direct current (DC), and the peaks, zero crossings, and polarity reversals of alternating current (AC). These drawn depictions will be recognized as similar to the oscilloscope pictures described in Chapter 2.

It will be noticed that alternating current or voltage has a peak level higher than the RMS (**Root Mean Square**), or power-effective level described on page 6. This is because there are regular falls to zero between the peaks as the polarity reverses. RMS values are fully described on pages 36 to 39.

The waveform shown is **Sine Wave**, the simplest alternating wave. In the many departments of Audio, sine wave is encountered frequently because it is the pure form of a single tone, without added harmonics. The nearest thing in nature to a pure single frequency tone is the sound of a flute, without the breathy overtones.

The majority of simple audio measurements, which will be introduced in Chapter 4, are made with sine wave, since its very purity of tone makes it possible to detect errors in audio signals; effects like noise and distortion that should not be present.

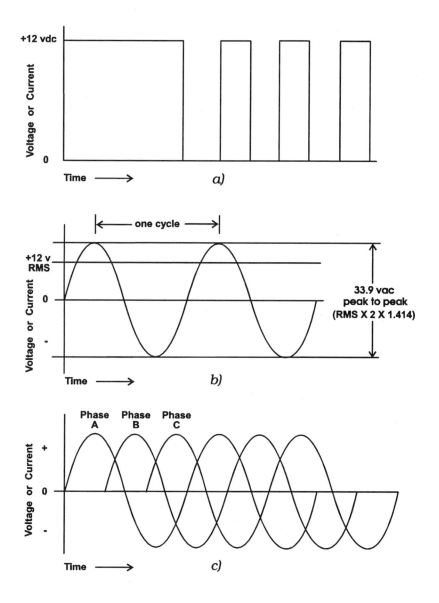

Fig. 1.5 a) Direct current or voltage, constant and pulsed.
b) Alternating current or voltage, and the relationship between RMS
and peak, as detailed in Chapter 3 and Chapter 11.
c) The combined AC waveforms of a three-phase mains supply,
showing the phases 120 degrees apart.

Chapter 2.

Meters and Electrical

Measurements

MULTIMETERS

While there are many specialized measuring instruments, routine measurement of Volts, Amps, Ohms, and their fractions, is conveniently done with a multimeter, ranging in value from twenty dollars for a pocket meter, up to several hundred dollars for a portable multimeter of high accuracy.

The choice of a digital or analog meter depends on the functions being measured, and to some extent on the accuracy required. AC volts and amps ranges should read as close as possible to true **RMS**, a waveform related convention detailed in Chapter 3, as the accuracy of calculations depends on initial measurements, especially calculations like watts where the voltage measured will be squared, and where a small error in measurement becomes a much larger error in the final calculation.

Analog meters are generally more fragile than digital, as the pointer is traditionally mounted in needle bearings and it won't survive a fall, but a rugged **Taut Band Suspension** meter movement is produced by some manufacturers. It can be well worth the relatively small extra cost.

Digital meters, subject to certain provisions, have higher accuracy than the cheaper types of analog meters, but an analog display is essential for observing rapid variations in voltage, like audio program, the real sound signal in lines and amplifiers. In this case, the ballistic characteristic of the meter movement needs to be known, so that voltage variations faster than the standard mechanical meter will follow can be predicted (see Reading a VU Meter, in Chapter 8). Some multimeters have an approximate dB scale.

Digital meters present a readout with high accuracy to several decimal places, which is very handy when charting routine measurements. An important point in choosing a digital meter is to select one with at least one more digit below zero than you wish to read with accuracy. The last digit may not be accurate, but will indicate that the second last digit can be relied on. Similarly, choose an analog meter with an ohms x 1 scale. You'll need it for working with speakers.

Meter ranges can include very low values. Simple analog meters, which are essentially electric motors that rotate through an angle against a return spring, do not have the sensitivity to read small quantities like millivolts and microvolts, milliamps and microamps, and resistances of several megohms. Millivoltmeters have DC amplifiers which provide the necessary sensitivity, and readings are then displayed in the conventional manner.

A digital meter may have millivolt ranges at almost no extra cost, because it is already an active device. If it is necessary to use an analog meter because the signal is varying, then the AC millivoltmeter can be expensive and often will not include a host of other useful functions.

Amplified meters present a high impedance input to the circuit under test, and do not modify the quantities being measured to any extent.

The following simple descriptions of meter circuits and methods will be useful as a starter for either beginners, or people already working in other areas of Audio Vision and Electronics, but who find that they are handicapped by lack of familiarity with meters and measurements and would like to use them or just expand their knowledge.

These descriptions are very basic, but as in almost every other aspect of training it is not the book that pumps information into the student, but it is the student who is self-taught with the book as a guide. The people that realise their ambitions in this field are the ones who first learn the fundamentals, and then advance their own knowledge through hands-on experience.

VOLTAGE MEASUREMENT

Volts, millivolts, and microvolts

Voltage is electrical pressure; it can be measured either across the source or across a load. In either case we are measuring **voltage drop**, or **potential difference**. Fig. 2.1 shows that voltage can always be measured without opening the circuit to insert test equipment; the voltmeter need only be bridged across the measurement points. Although AC ranges read voltage which presents itself alternately in both directions, DC can not normally be accurately read on an AC meter range with the average multimeter. Multimeters read the RMS value of AC voltages (page 36), because that is what we mostly need to know.

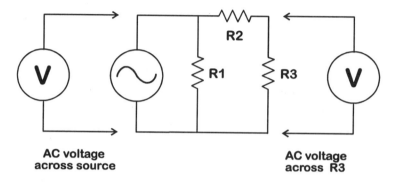

AC voltage
across source

AC voltage
across R3

Fig. 2.1 Voltage measurement across components.

AC or DC volts is selected as required, the meter is switched to a voltage range giving a full scale reading higher than that anticipated, and the meter probes applied (positive to positive) across the part of the circuit to be measured.

Note that the two readings of Fig 2.1 will be different; voltage drop across R1 will be equal to the source voltage, and the sum of the voltage drops of R2 and R3 will equal the source voltage, but the measurement across R3 will be less than the source voltage, unless R2 is zero ohms.

Because voltage readings so often involve a network, like the one in Fig. 2.1, it is always necessary to think of voltage as a **difference** quantity; as **Voltage Drop**. This is just as valid for single voltage sources, like power supply or battery outputs, but the concept of voltage drop will considerably assist understanding of voltage in more complex circuits.

VOLTAGE DROP IS MEASURED ACROSS COMPONENTS OF A CIRCUIT

CURRENT MEASUREMENT

Amps, milliamps, microamps

In its simplest form, an ammeter is a low value series shunt resistor sampled by a voltmeter (Fig. 2.2). The shunt value is a fraction of an ohm, low enough not to seriously affect the circuit under test. While voltage measurement is made across components in a circuit, current is measured in series with them in order to intercept the flow. However, current can also be calculated from voltage measured across one circuit resistance element. Like voltmeters, ammeters measure RMS current where AC is concerned.

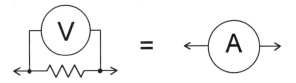

Fig. 2.2 An ammeter measures volts across a shunt resistor.

The voltmeter method of circuit current measurement is the basis of **current sensing** as used in many electronic systems, for example, some amplifier overload protection and negative feedback applications.

As with voltage measurement, the meter is switched to a safe Alternating or Direct Current range before the current is applied to it.

It is worth repeating that the current through a single un-branched circuit is the same at any point in that circuit. Referring to Fig. 2.3, the current flowing through R2 is the same as through R3. However, if the resistances are different, then it follows that for the same current, the voltage drop across R2 and R3 will also be different.

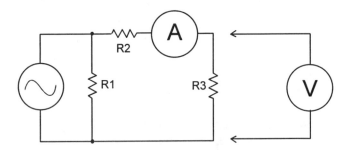

Fig. 2.3 Current measurement through R2 and R3.

**CURRENT IS MEASURED IN SERIES WITH A CIRCUIT
OR CALCULATED FROM THE VOLTAGE DROP
ACROSS A SERIES RESISTOR**

When making measurements for the purpose of determining watts in a known circuit resistance by calculation, as in amplifier output measurement, it would be just as accurate to measure current instead of voltage, and calculate watts from the two known quantities. However, current measurement involves opening the circuit to insert the meter, whereas voltage measurement does not, so for the sake of simplicity, calculations like this are made from voltage measurements, all other things being equal.

A WORD OF WARNING ABOUT METER CURRENT RANGES

While selecting the wrong voltage range may result in the meter needle being swung violently against its stop, selecting the wrong function, like milliamps for voltage measurement, can result in high currents passing through the current shunt and meter movement, with consequent damage. For example, a dry battery is capable of delivering several amps for long enough to damage the meter.

METER CURRENT RANGES SHOULD NOT BE APPLIED TO A VOLTAGE SOURCE

RESISTANCE MEASUREMENT

An ohmmeter is a voltmeter with an appropriate resistive network and a voltage source (see Fig. 2.4). The network usually consists of switchable shunt and series resistors on the voltmeter to produce the usable ranges, for example: Ohms times one, Ohms times one hundred, and Megohms (millions of ohms). Its a good idea to compare the meter's readings against a known standard, and before purchasing it, verify that it is capable of being calibrated and maintaining its accuracy.

There are other instruments for measuring resistance, including resistance bridges for high precision, but a good multimeter goes a long way to serving the needs of the field and workshop technician.

RESISTANCE IS MEASURED ACROSS THE CONDUCTOR (By passing a small current through it)

Note that a resistor under test should be isolated from the rest of its circuit by disconnecting one end, if the rest of the circuit in parallel with it is going to make a difference to the reading, but in some cases, where

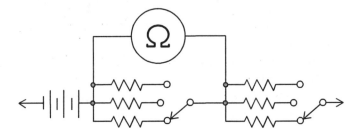

Fig. 2.4 An ohmmeter is a resistive network with a voltage source, sampled by a voltmeter.

continuity verification or an approximate measurement is required, then it may be satisfactory to make the measurement with the component fully connected. Diode short circuits can be found without disconnecting them, if measured in both directions.

Another factor in measuring diodes, polarized capacitors, and the like, is that most ohmmeters present negative volts from the internal battery at the positive probe, and that this polarity anomaly can be verified with another meter before measuring polarity conscious components or circuits.

USING ANALOG AND DIGITAL MULTIMETERS

Both types of meter are illustrated in Figs. 2.5 and 2.6. They are simple meters, and until the reader is more at home with them, it is advisable to avoid meters with a lot of extra functions. Auto-ranging is a very handy facility, but again, it can be totally confusing as one can get lost, not knowing which range it has selected. Buy cheap, simple meters first, and by the time you need to invest in an expensive one, you will know which to choose.

ACCURACY OF A METER

When making measurements, there are occasions where precision is not a priority; for example, if two voltages are to be compared, or an approximation will suffice. But there are also times when accuracy is vital, because other quantities may be derived from the measurement, or it may be required to balance a voltage or current against a published figure of a critical nature. Two classic examples that come to mind concern amplifier alignment:

Fig. 2.5 The Analog Meter.

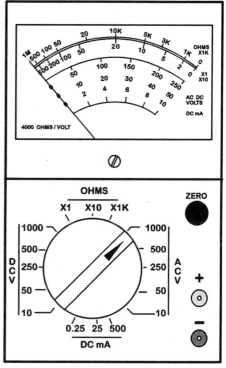

The four functions; AC volts, DC volts, Ohms, and DC milliamps, are all divided into convenient ranges. The screw under the scale mechanically zeros the pointer when it is at rest. It does not need regular attention.

Before making ohms measurements, hold the probe tips together and zero the pointer with the knob. If it won't reach full scale, replace the battery. Use the appropriate scale to read ohms. Don't apply a voltage to the OHMS or DC mA current ranges. It will damage the meter.

When measuring volts, select AC or DC first, then use the range that will read above the test voltage to protect the meter. If you need to, you can then switch to a lower range during the measurement. Apply the positive probe to positive voltage, and so on. Read from the appropriate scale whose highest figure represents the range selected. Remember to check that the probes are in the correct polarity sockets before using the meter for any function.

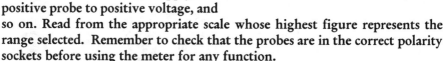

After use, select the highest AC volts range, as the battery will be drained if the probes remain in contact on OHMS. Some meters have an OFF position which shorts the moving coil to damp it.

1. AC output voltage of an amplifier under test is squared, then divided by the load resistance in ohms. The resultant figure in watts will be inaccurate by many times any error of the voltage reading, because the voltage was multiplied by itself.

2. Amplifiers sometimes require re-balancing of **bias** and **DC offset voltages**, especially if components have been replaced. Although outside the scope of this book, Service Manuals give the specification, how and where to measure it, and how to adjust it. They are particularly important adjustments because they can make the difference between the amplifier continuing to function reliably or suffering damage. Only advanced technicians should set bias.

Fig. 2.6 The Digital Meter.

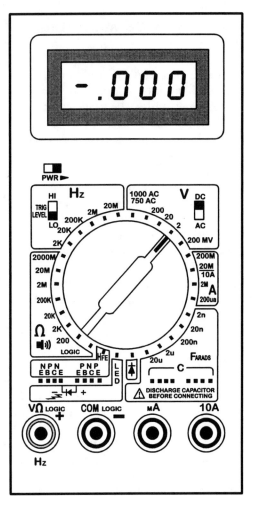

This one does much the same as the analog, but has a lot of handy extra functions. It's relatively easy to add powered functions to a digital meter because it already contains active circuitry. These include LED and transistor testers and capacity measurement.

The display has three digits to the right of the decimal point, indicating that it will be accurate to two decimal places.

Otherwise, it is straightforward to use, and has greater accuracy and more ranges than the analog. Remember that the last digit on the display is mainly there to ensure you can depend on the second last for accuracy.

Special probes and leads are available for safety when measuring high voltages. If high currents are measured on, say, a ten amp range, use heavy gauge leads, and limit the exposure time to let the internal shunt cool down, to protect the meter from excessive heat. External shunts can be used for higher currents, measuring it with a low volts range from which to calculate amps.

Both adjustments depend on accurate measurements because the voltages in question must be set according to a published specification. The other point about accuracy is that a major factor in AC readings is the meter's ability to read **true RMS**. AC measurements are meaningless if the relationship of the measurement to peak value is unsure. The RMS (root mean square) value of sine wave alternating current is 0.707 of waveform peak value.

The term peak has two different meanings. A peak voltage may be the maximum value achieved in a particular long term exercise, but waveform peak is the maximum excursion of the wave, either positive or negative, within any one cycle. RMS and peak values are described in detail in Chapter 3.

METER SENSITIVITY

When choosing an analog meter, one of the features for which it should be selected is its **sensitivity**, which is referred to in ohms per volt. The intent of measuring is to obtain the reading without loading or otherwise modifying the circuit under test. An ideal voltmeter will present to the circuit as high a resistance as practical; an ideal current measuring shunt will have a resistance close to zero, in comparison with the impedance of the circuit under test.

Try this simple test with a cheap analog meter and a flashlight bulb: Select Ohms x 1 and measure the resistance at about 5 ohms, but note that the needle continues to drift slowly. The lamp filament is getting hot due to the meter current passing through it and therefore its resistance is rising. Tungsten filament lamps of this type are typically 5 ohms cold, and 50 ohms at operating temperature. Like most conductors they have a Positive Temperature Coefficient. However, meters that pass appreciable current are useful for checking things like diodes and speakers; most silicon diodes have higher forward resistance at low currents, and give uncertain readings, and speakers will not audibly verify continuity if checked with high impedance meters, so most digital meters have a diode test position (page 24).

A useful installation tip concerns identifying both ends of multiple long lines. A short circuit applied to the far end will identify the line by default at the metering end, provided there are no other short circuits, but a diode connected at the far end will positively identify the line because it will read either as a short circuit or as an open circuit, depending on the applied meter polarity. An ohmmeter with in-built buzzer is the ideal tool for continuity testing, but choose one with a fast response beeper; it will save you a lot of time. A set of universal probes is also very handy, enabling the user to either prod a sharp tip onto tarnished or lacquered connections and penetrate the layer to make an effective connection, or to clip on to a conductor in crowded areas of circuit boards like a third hand.

ADDITIONAL FEATURES AVAILABLE IN MULTIMETERS

Many multimeters come with useful features like transistor testing, capacity measurement, and frequency counter, but in a compact portable device these may only be very basic measuring tools, for example, the frequency counter may only lock on to very pure waveforms, giving erratic readings if harmonics are present. When choosing a meter, extra features that may detract from the primary functions of Voltage, Current, and Resistance measurement, should be viewed with some reserve, unless they are particularly needed, and their shortcomings understood and accepted.

Finally, anyone who owns a high precision analog meter will be advised to carry a cheap one on field work, provided it will do the job, and leave the expensive one in the workshop, where it can be calibrated by one of the many organisations offering this service, and periodically used to double check the accuracy of the meter that takes the risks. In any event, some precision analog and digital meters are best left on the test bench for the duration. More rugged types like most hand digital meters, and analog meters with "taut band suspension," are inherently able to sustain the shocks encountered in everyday use.

Although many instruments come with carry cases, its a good idea to organize one big enough to contain the meter fully set up, ready to use, complete with probes connected, folded beside the meter, and enough room for alternate probe clips or other accessories already in place. Much equipment comes "Gift Wrapped" so neatly that it won't go back in its box without fully dismantling the leads and neatly rolling them up. The astute technician will re-package portable equipment so that it requires no preparation other than to open the case, pick up the probes, and start using it.

OSCILLOSCOPES

The ability to see voltages and waveforms in graphic form is a valuable adjunct to test procedures, and although not essential for audio work, owning at least a basic oscilloscope as a learning tool is recommended. Originally known as a Cathode Ray Oscilloscope, or CRO, the oscilloscope is also made as a digitally processed liquid crystal display, so Oscilloscope is the appropriate description. In fact, the ultimate technician's tool available today is a portable multimeter with a liquid crystal waveform display.

Single or dual-trace, for audio alone or for TV servicing and observing digital signals, 100 kHz or 100 mHz bandwidth, analog or digital, the choice of an oscilloscope will involve an outlay ranging from a couple of hundred dollars to several thousand.

Chapter 3 describes the use of oscilloscopes for refining amplifier output measurement. A relatively low cost one will do that, and it does not have to be particularly portable since field measurements can be accurate enough without it. Another function of simple oscilloscopes is distortion analysis.

While it is possible to gain a rough assessment of distortion by observing the waveform of a steady tone, the X-Y observation is more precise, and is a useful diagnostic and final checking procedure. Signal comparison can be made by sampling the input signal of an amplifier at the X-, or horizontal input, and the amplifier output signal at the Y-input. The oscilloscope resolves the signals into a diagonal straight-line display.

The vertical input is accessed via the external sync input sockets and the Sync Mode switch. Adjusting the relative gains of X and Y channels will set the displayed line at 45 degrees, and any difference between the two signal sample will be shown as deviation from a straight line, whether the difference is Harmonic Distortion, Transient Intermodulation Distortion, Crossover Distortion at the output stage, or an entire range of ills that call for search and diagnostic observation.

A budget model (Fig. 2.7) will enable you to make many interesting observations, including calibrating the screen to make peak voltage measurements directly from the display image. If the display is linear, RMS can be calculated from peak measurements by using the 0.707 RMS to Peak ratio.

An expensive one will enable advanced technicians to set up CD players, examine modulated carrier waves, and capture waveforms for study with the sample and hold facility. Most oscilloscopes available today only have one discreet beam, and dual or multiple trace units employ multiplexing to enable the user to see, for example, several related digital signals side by side. It is difficult at first to know what to buy until sufficient hands-on experience or training indicates which types of oscilloscope have the features for your intended specialty. Unless there is a specific need for an expensive oscilloscope, the recommendation is that a two hundred odd dollar outlay will provide an informative and useful experimenter's tool that will assist in troubleshooting and speed up outine measurements in the workshop. Oscilloscopes are among the less used tools a technician owns, so like multimeters, get a cheap one first unless you particularly need an advanced model.

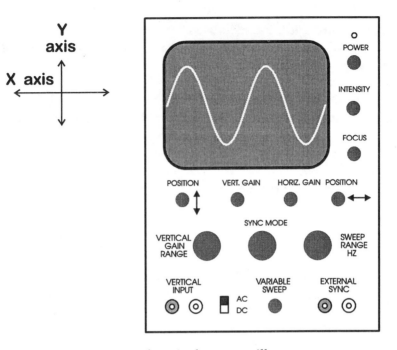

Fig. 2.7 **Budget single-trace oscilloscope.**

OSCILLOSCOPE EXPERIMENTS

Use a small photo-voltaic cell (solar cell) to view the output of an infra-red remote control. The cell is a self-generator in the presence of visible or infra-red light, unlike a photo-diode, and needs no other power source or bias. Different remote transmitters will be seen to have a different number of pulses in their "words," from seven for early VCRs to about thirty, and the pattern will be visibly different for each function.

Feeding the X input with low voltage at mains frequency, and the Y input with the output of a variable frequency audio signal generator, will reveal the classic "Lissajous" figures, which display the phase relationship etween signals of the same frequency and different frequencies. Modulating the two signals would be capable of producing similar trace drawings to laser projectors, which use X and Y servo driven mirrors.

Chapter 3.

Amplifiers and Transmission Lines

POWER AMPLIFIERS AND LOADS

A line level audio signal will not directly drive a speaker because it cannot deliver the current required to maintain its voltage across a low impedance load. Speakers look like short circuits to line signals, even though sufficient voltage for low level reproduction may be present before the load is applied. So the power amplifier is inserted between the line source and the load to make current available as the signal voltage directs.

A power amplifier is a current converter, and the output has a source impedance much lower than its load. The power supply has high value storage capacitors which give it a low source impedance too, and it delivers current through the output stage, controlled by the input signal voltage. Source impedances are therefore intentionally low so that the output voltage will not fall under load.

Amplifiers also use negative feedback to preserve a true **transfer characteristic,** or so that the output will be true to the input; the more the load drains it, the harder the amplifier pushes, up to the limit of its ability. Amplifiers normally have enough gain to multiply a line level of about one volt up to the voltage representing full power output.

Gain is **signal multiplication** or amplification factor, and should not be confused with **Power Output**, a different quantity which is **the amount of energy able to be delivered to the load** at acceptable distortion. The loose term **Volume** basically means acoustic level, or loudness, and is not the same as gain or power. Professional power amplifiers do not have high gain, just enough to bring line level up to full output. Domestic amplifiers usually have more gain to enable low level sources to be brought up to normal levels. A loss involving a few decibels at line level is relatively easy to recover just by turning up the gain or adding a small recovery line amplifier, but use of resistance elements in a speaker line represents power lost in heat, and should not be considered as a volume control or impedance matching method if there is any other conceivable way around the problem.

Power amplifiers can deliver more energy than their rated output, but it is at the cost of distortion, for as output increases beyond the rated maximum, the available voltage limit clips the top off the signal peaks, and waveform changes progressively from sine to square wave. Although the peak output voltage can not increase beyond that available from the power supply, the duty cycle of the waveform increases, raising the total energy output in a form damaging to speakers (Fig. 6.1).

The RMS voltage from which amplifier output power is calculated was defined on page 24, and when **RMS** and **clipping** are detailed later on pages 36 and 40, it will be seen that as an increase in clipping forces the Mean Value of the signal up towards Peak Value, up to 30 percent increase in output can occur, but in a form that damages speakers, since square wave has a similar effect to Direct Current on speaker voice coils (page 100).

GAIN IS SIGNAL MULTIPLICATION, IN DECIBELS
OR A NUMERIC MULTIPLIER

OUTPUT POWER IS ENERGY DELIVERED
TO THE LOAD, IN WATTS

THE INPUT CIRCUIT

The input impedance of an amplifier is generally made low enough to maintain a termination at all times for reasons of low noise and stability, and high enough to maintain a bridging condition, even if more than one amplifier input is connected to a signal source (Fig. 3.1).

Most power amplifiers have bridge inputs. Bridging is a condition where the input connected to a source has an impedance sufficiently high that it will not alter the performance of an already loaded source. An input with an impedance five times the minimum line termination is considered to be a bridge input. Where necessary to produce a bridging condition, series resistance can be added to the input circuit. This will introduce a loss, and making a bridge input this way has to be considered against the availability of gain to replace the loss, when planning a system. The use of bridging transformers to match the required impedance with minimal loss is an efficient but expensive alternative; input transformers have to be magnetically shielded, suitably positioned, and of good quality to avoid induced noise and signal degradation.

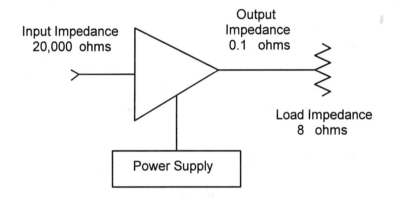

Fig. 3.1 **Elements of a power amplifier.**

Inputs sometimes require a terminating resistor to reduce noise and stabilize a system or when measuring noise level. Source impedance is not the same as load impedance; it can be between 20 and 100 times lower. When speaking of circuit impedance, we refer to the load or termination impedance.

THE OUTPUT CIRCUIT

The output impedance of an amplifier is quite a lot lower than its load impedance. Speakers need to be closely coupled to the amplifier to take advantage of the low source impedance which compels them to follow the waveform. Cable losses spoil this coupling, and also waste output power in heat. Choice of speaker cable is an important influence on **Damping Factor,** which governs the hold the amplifier maintains on speaker motion. The power loss due to cable resistance is a leading factor in amplifier use; an 8 ohm speaker connected with a cable whose total resistance (both conductors) is half an ohm will waste one 16th of its power; in a 50 watt power output, only 47 watts will be available at the speaker.

The resistance of the cable chosen will depend on its length and its conductor cross section area, so one concept of a low-loss line is based on the approximation of one square millimeter conductor cross section for every 10 meters at 100 watts. For example, use 3.5 square millimeter cable for 10 meters at 350 watts or 35 meters at 100 watts. Although compromise is usually acceptable, attempts should be made to keep losses to a minimum.

Amplifiers can therefore be located close to the speaker for optimum performance, but this in no way suggests it is mandatory. In the majority of situations there are conditions of convenience that indicate central amplifiers and distributed speaker lines. In both professional and domestic fields, separate amplifiers and speakers predominate over powered speaker systems, despite the inherent advantages of the latter.

AVOIDABLE FAILURES

Speaker lines are high current circuits, whether they drive transformers or speakers directly. While many of the following items are desirable, each one is a candidate for future losses or breakdown. A speaker line could conceivably contain some or all of the following: Amplifier output connector, the solder joints on the back of the connector, speaker line relay contacts and the screw terminal or solder joints in and out of the relay, screw terminals in the speaker line connection box at the other end where the line might change from solid to flexible conductors, fuses, fuseholders and their solder joints, and the speaker input connector's solder joints and the contact resistance of the connector pins. If both conductors are counted, that adds up to twenty points where aging solder joints, loose screws, or oxidized contacts can add resistance to the circuit, or just fail when the weather dictates. Add to this the connections inside the amplifier and speaker.

All these possible failures are mentioned, because they really do happen. It is obvious that a compromise has to be chosen between including useful items and just plain overloading of the system with failure factors. A speaker line is a low impedance device, it has low voltage and high current, and looking out for stray ohms is an important part of a no-fail installation with a long life expectancy.

A discussion of output circuits would not be complete without mentioning that speakers are not the only loads that can be driven by amplifiers. Watts are watts, no matter whether we use them to make sound, or drive motors. Like other forms of energy, watts can be converted into any other form of energy; sound, heat, motion, light, and so on.

Square and sine wave inverters have output stages similar to audio amplifiers, even if they have a different style. Some audio amplifiers use switching output stages to produce **PWM** (pulse width modulated) outputs to drive speakers, like a modulated power inverter. Amplifiers have long been used to drive electric motors in control systems, and many of the parameters are the same; power output, frequency response, and stability. The knowledge that Servo Systems drive motors with amplifiers assists our understanding of speaker amplifiers and negative feedback even if we never become involved with servo systems.

SPEAKER IMPEDANCE MATCHING

Some professional amplifiers will tolerate loads of one ohm or less, so that, for example, eight speakers, each of 8 ohm impedance, can be connected in parallel. But it is important to remember voltage drop in the speaker line, and to connect each speaker with its own cable directly to the amplifier, instead of feeding a cable from the amplifier and looping through from speaker to speaker. By the latter method, the last speaker in the line will receive the least power, and the cable from the amplifier to the first speaker will introduce the greatest loss or get hot unless it is suitably heavy cable.

Most domestic amplifiers over 80 watts will tolerate a 4 ohm load unless otherwise specified, which means that two 8 ohm speakers can be used in parallel. If more speakers are needed, a series-parallel network will provide almost any desired impedance, but because speakers are generators as well as being motors, it is important not to connect non-identical speakers in series, because they will not share the signal evenly at all frequencies and will tend to drive each other at various parts of the spectrum, leading to strange effects. Fig. 3.2 represents three series speaker strings in parallel. If each speaker is 8 ohms, then we have three 24 ohm strings in parallel; making a total of 8 ohms.

Notice in this calculation that the number of identical speakers in a **square format** series-parallel network like the one shown in Fig. 3.2, **whose square root is a whole number** (4, 9, 16, 25 speakers, etc.), will always have a total impedance equal to that of one speaker. The following impedance formulae are for identical speakers, resistors, or inductors:

*SPEAKERS IN PARALLEL: DIVIDE THE IMPEDANCE OF
ONE SPEAKER BY THE NUMBER OF SPEAKERS.*

*SPEAKERS IN SERIES: MULTIPLY THE IMPEDANCE OF
ONE SPEAKER BY THE NUMBER OF SPEAKERS.*

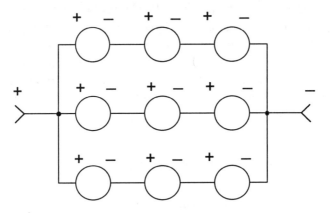

Fig. 3.2 Speakers in series-parallel.

CONSTANT VOLTAGE LINES

When driving a large number of speakers, especially over a great distance, it is convenient to employ **Constant Volt Lines**. The voltage is not exactly constant, as the name might imply, but rather it is a system operating on a standard RMS line voltage representing maximum amplifier output. 70 volt and 100 volt systems are common, but higher voltage lines, like 140 volts, are used for larger amplifiers.

Constant voltage lines have two big advantages. Like power lines, high voltage speaker lines can use smaller cross section conductors because the current is lower for the same number of watts. Secondly, they use a step-down transformer at each speaker, so all speakers are connected in parallel. This is a major advantage, because failure of one speaker will have minimal or no effect on the others, unlike a series network where open circuit of one component will shut down all others in that string. Another advantage is choice of output level at each speaker by means of switchable tappings on the speaker transformers. While most speaker transformers present a primary impedance to the line between 200 and 500 ohms, it is practical to use a 10,000 ohm transformer in an area like a washroom or a quiet office, where low volume is the requirement for background music and announcements. The system is conservative of amplifier power; each speaker takes only the power it needs.

The constant volt principle has been in use for a long time, since the days when amplifiers used output transformers. Then it was a matter of selecting the right number of secondary turns to select between 2 and 25 ohms for speakers, or 70 volt line. Today we use amplifiers with transformerless output stages, and most amplifiers are either two channel or stereo (not exactly the same thing), so there are multiple choices available for line driving. As a general rule, stereo amplifier outputs can not be arbitrarily interconnected, but many two channel amplifiers are designed with switchable bridge or parallel mono configuration.

Bridge mono effectively places the two outputs in series, with the output taken from the two positive terminals and no connection made to the negatives, one input phase being inverted so that the two inputs sum. This results in more than double the output in watts available from a single channel because the voltages and impedances are added.

If the power output into a nominal load is such that 65 or 70 volts is available, then the amplifier will drive a 70 volt line without using a step-up transformer. The total output watts available is then distributed among the parallel connected, transformer coupled speakers on the line (Fig. 3.3).

It is fortunate that high output amplifiers do not need a line transformer on the occasion when it would be a large and expensive one. However, if the amplifier can not provide an output close to line voltage, then a transformer will be required. Both isolating and auto-transformers are used.

When a transformer is closely coupled to an amplifier of low source impedance, it is not the same type of load as a speaker. Speakers have very close to the same DC resistance as their nominal impedance. Transformers, on the other hand, are among the most efficient devices, they have a

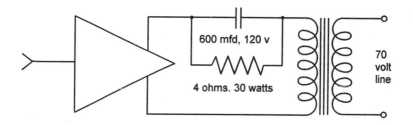

Fig. 3.3 Coupling an amplifier to a line transformer.

higher Q, as is said concerning the ratio of inductive or capacitive reactance to resistance, and present a very low DC resistance to the amplifier. Therefore, to protect the amplifier and transformer from high level, low frequency transients, it is necessary to include a high pass filter to decrease the coupling at near DC frequencies, as shown in Fig. 3.3. The capacitor attenuates extreme low frequencies, but has minimal effect on normal program. Inclusion of this network is a good idea also when there is no transformer at the amplifier end of the line, but when there is one on each speaker, because a load consisting of several small transformers will similarly have a very low impedance at near DC frequencies.

The resistor should be rated not less than one tenth of the amplifier's maximum output, or ten watts for every 100 watts of amplifier power. The capacitor should be chosen for its alternating current carrying capacity, since the peak voltage across it is not the only factor of importance. In this application, high transient currents are expected, so a polyester capacitor or **motor start capacitor** is in order. A capacitor working voltage as low as 120 volts AC is suitable, as its safe peak voltage will be higher. Motor start capacitors are available from refrigeration spares suppliers. **Non-polarized electrolytic capacitors** are not suitable for this application, as they will not survive the current peaks, and their use is only recommended for domestic crossover networks of moderate power.

RMS VOLTAGE AND CURRENT

In alternating current systems, including audio power circuits operating at constant amplitude and frequency for the purpose of measurement and discussion, watts is the product of volts and amps that rise to a peak and fall to zero at twice the frequency rate, as they swing positive and negative each cycle.

The resultant power, provided the volts and amps are in phase, is not calculated from peak voltage or current, as the peak is a short dura-

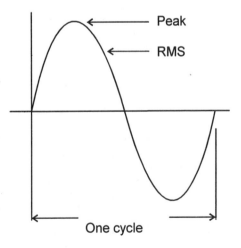

Fig. 3.4 **Elements of a sine wave.**

tion event, and a sine wave has intervals when voltage or current falls to zero as it crosses to the opposite polarity. Instead, watts is related to the **RMS** value of volts and amps, which is 0.707 of the peak value of a sine wave (Fig. 3.4).

RMS means **Root Mean Square**, and is the form normally used for Alternating Current and Voltage measurement. Mains voltages are quoted in RMS values, and AC ammeters and voltmeters read RMS. Unless otherwise stated, all references to alternating current or voltage refer to the RMS value— **the part that does the work** when the effect of all the peaks, troughs, and alternations are combined. Peak voltage is referred to when assessing the point in the waveform at which clipping occurs, but only RMS tells us about the part of the waveform that does the work. RMS is not **Average** current or voltage, but the **Geometric Mean**, or 0.707 of peak value.

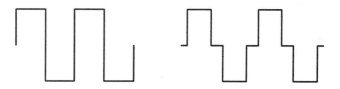

Fig. 3.5 **Square wave and modified square wave.**

There are two other waveforms that should be introduced here: **square wave** and **modified square wave** (Fig. 3.5). The latter has an RMS value like sine wave. Normally, audio can be said to be made up of little bits of sine wave, depending on the frequency or angle of the waveform at any given instant. But waveform **clipping** occurs in amplifier overload situations, which in extremes produces a square wave in which the smooth transitions of sinusoidal waveform are replaced by instantaneous switching from one polarity to the other at twice the fundamental frequency. Because of the sharp angles at each polarity shift, denoting rapid change, square wave is a multi-frequency waveform, reaching into the megacycle band and beyond, if the system permits it. Some power inverters generate square wave, the simplest way to make AC from DC, but because there are no intermediate voltage or current levels in square wave, there is no RMS value, only peak.

Electric motors, mentioned because they are very like speakers, and help to demonstrate the value of RMS, are wasteful convertors of square wave energy into motion because they are unable to follow the peak to peak polarity changes, and they overheat, among other ills, so suitable power inverters produce either sine wave or modified square wave which gives the RMS and peak values the correct relationship of 0.707 to one. Further clarification of RMS is given in Chapter 11 in the section on **Rectifiers.**

*THE RMS VALUE OF A.C. VOLTS AND AMPS
IS 0.707 OF PEAK VALUE*

Peak is the reciprocal of 0.707; RMS times 1.414.

POWER

Current and voltage can be referred to as either RMS or Peak, but the terms have no relevance to watts since the watt represents **Effective Power,** or heating effect, and **only exists as the product of RMS quantities.**

While it was said in Chapter 1 that watts equal volts times amps, there are conditions attached where AC is concerned. For example, transformers are often rated in **volt-amps** instead of **watts** because circuit efficiency depends on the **relative phase** of voltage and current. VA (volt-amps)

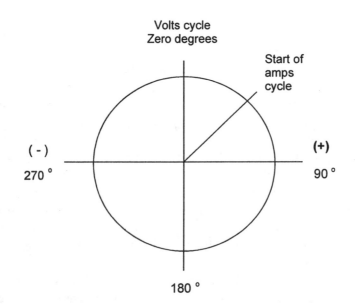

Fig. 3.6 Voltage leads current in this 360 degree display of an AC cycle.

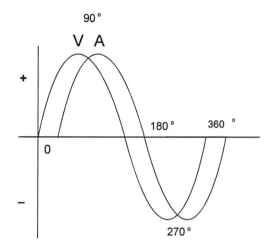

Fig. 3.7 One voltage cycle showing amps lagging by 45 degrees.

equals watts only at circuit **Resonance** (to the power frequency), or at **Unity Power Factor**, when volts and amps are **in phase**. Figs. 3.6 and 3.7 show one cycle of AC in two forms; as a 360 degree voltage cycle, and as voltage and current sine waves.

The accompanying current wave is shown starting late, in each diagram; volts are leading, amps are lagging. Where one waveform leads or lags the other due to circuit inductance or capacitance, watts no longer equals volts times amps, but is less by a calculable circuit efficiency rating called **Power Factor,** in electricity distribution parlance. It is not relevant in normal speaker lines.

Diagram convention displays waveform elapsed time from left to right, and cycles, clockwise from the top. It can be confusing and ambiguous reading these diagrams; which is the leading component ?

The simplest solution is to imagine a cursor moving clockwise from the top, or from left to right in the case of the linear graph in Fig. 3.7. The first curve it touches is leading.

AMPLIFIER OUTPUT POWER MEASUREMENT

An amplifier can be regarded in this analysis as a device with an input, an output, and a power source. Most of what we need to know about its output power can be measured with a signal generator and a voltmeter across a known load resistance. As shown in Fig. 3.8a, a high wattage resistor is used for the amplifier load because it is frequency independent and generates no back-EMF.

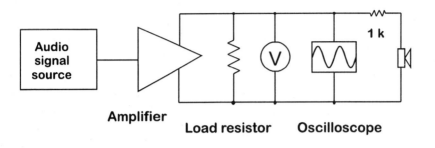

Amplifier   **Load resistor**   **Oscilloscope**

Fig. 3.8a Test setup for output power measurement.

The oscilloscope, speaker monitor, and voltmeter make it possible to see, hear, and measure the test signal across the output load. The signal generator (or AF oscillator), can be a simple one, but should be capable of producing a sine wave that looks clean on the oscilloscope, or is sufficiently low in distortion to allow an obvious indication to appear when the amplifier starts to clip the waveform as it exceeds maximum rated output.

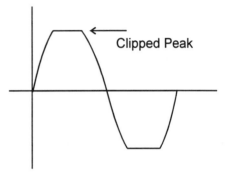

Fig. 3.8b Test Signal clipping.

Precise distortion measurement is not the aim. A low cost battery portable signal generator for amplifier field testing will be a handy adjunct to the dummy load described on page 44.

In order to assess the amplifier's maximum output, we need to know when clipping occurs, so the level of a test tone between 600 Hz and 1 kHz is advanced slowly until it is reached.

Fig. 3.8b shows how the peaks of the waveform are clipped as maximum output is exceeded. The test operator should ensure the amplifier is not clipping earlier; perhaps at the input stage, by making another test with the amplifier's gain control fully advanced, and the tone source brought in slowly from a lower level. Clipping occurring before the output stage will not permit measurement of an amplifier's output capability because maximum output will not be reached.

When the waveform peaks are flat-topped, instead of one peak, there are now two. So as well as the fundamental frequency there is now a second harmonic content present (the second harmonic is twice the fundamental frequency). As discussed at the beginning of this chapter, the amplifier can give more output than its rating, but it is the **undistorted output** we need, **just before clipping**.

Harmonic distortion is addition of harmonics. In a precise measurement, onset of clipping could be detected by observing the harmonics on a wave analyser or a high resolution audio spectrum analyzer, but observation of clipping on an oscilloscope is satisfactory. These methods have the advantage of being usable at any frequency, although it is generally inadvisable to hold an amplifier at full output for more than a few seconds at very high or low frequencies. Accurate enough for most purposes, a simple way of finding the clip point in the field or in a quick check is to select a frequency of 1 kHz or less, low enough to have an audible second harmonic. A 2 kHz harmonic will be easily detected by ear when it appears, as the signal level is increased to maximum output.

The level we want is just below clipping, and watts will be calculated from the RMS output voltage at that point. Measure it precisely, because we're going to square the voltage reading.

Some professional amplifiers have input-output comparators built in. They should be used to indicate onset of clipping in conjunction with other methods as they are more accurate than either of the methods given, and they also show up other imbalances, giving early warning of an amplifier about to fail. Reference to the power formula in Chapter 1 gives the power output:

$$W = \frac{V^2}{R}$$

A practical calculation example is given in Appendix A.

WATTS EQUALS VOLTS SQUARED
DIVIDED BY OHMS (Load)

DC OFFSET VOLTAGE

The following notes are not intended to be a guide for repairing amplifiers, but for detecting faults that require an experienced repairer, since further damage can result from continued operation where faults are evident. If an amplifier has been wet or seriously damaged, send it for service without turning it on.

Whenever output is measured, the DC offset voltage should also be checked. An amplifier's load expects to see an AC signal, but if the amplifier's DC balance is less than optimum for any reason, then DC may appear at the output along with the signal, or in the quiescent, or no signal state, and when this is measurable off load without a signal, it is likely to be very much more serious when the amplifier operates at normal levels into a speaker. Clipping of one peak before the opposite polarity peak also indicates a fault which should be fixed before the amplifier is used.

DC in the output will also be indicated by an input-output comparator, and the cause can be anything from maladjustment, which in a direct coupled amplifier can originate at the output driver stage or as far back as the input stage, or the source equipment driving it may be responsible.

Faulty output devices can also be the cause, and in some amplifiers these can be replaced individually. Other amplifier designs are safely repaired only if all output transistors are renewed together. Alternative types can not usually be mixed with original output devices, and an output stage crash is always expensive, so it should not be risked by allowing DC imbalance to continue.

DC bias, the fixed voltage applied to the inputs of the output stage devices, is not the **DC offset,** but is part of the setup for the driver and output transistors, and its purpose is to make them operate on the linear portion of their transfer curve (Chapter 5) so that the signal crosses over smoothly from one polarity to the other. Bias adjustment, where possible, is critical, and is carried out whenever output or driver transistors are replaced according to the service instructions for a particular amplifier. Although this book does not detail such adjustment to amplifiers, mention is made of the importance of these alignments which are specified in the service manual for each amplifier, so that the reader will be aware of the need to follow up amplifier testing by sending it for service where faults are indicated.

Some amplifiers have no DC offset or bias adjustments, especially those using integrated circuit output devices. So assessment and possible component replacement is in order. Following any replacement of components that will affect the offset, it should be measured in any case, as components vary. Measurement is simple, starting with the 50 volt DC meter range to sample the quiescent output, that is, with no signal at the output, then select lower voltage meter ranges until a reading is obtained. When in doubt, start high to protect the meter.

An obvious indication of trouble is a speaker cone polarised hard over to one side of its excursion, possibly preceded by a more that usually loud "thump" when the amplifier is turned on, and in this case, switch off before the speaker is damaged and disconnect it, select zero gain, and meter the output. Check also if the problem disappears when the amplifier input is disconnected, which might reveal that DC is coming from the source equipment, in the case of a DC coupled system. If a voice coil has been overheated due to this cause or through overload, it can often be saved by allowing a five minute cool-down period before using it again.

The safe amount of DC offset varies, depending on the load impedance and how hard it is intended to drive the amplifier towards clipping, but in general, a DC offset of 10 millivolts is acceptable, whereas much over 50 or 60 mV is getting too high for safety. An amplifier can still operate with a DC offset of half a volt, but to continue using it risks catastrophic failure, which may destroy the entire output stage, and also pass DC through the speakers or transformers, with the inevitable result of burn-out. From small beginnings, DC in the output can suddenly escalate into major expense.

CATASTROPHIC FAILURES

Recently an interstate acquaintance had a complete output circuit board assembly replaced in his 400 watt stereo amplifier. I have always advised people to go to a specialist when servicing big power amplifiers, because this is an often repeated story. Anyway, he elected to have the new board, complete with output devices and heat sink, fitted by a local service center. A few weeks later, after he had fallen asleep to music one night, he awoke at 6 am to find his amplifier disappearing in a cloud of smoke. Necessary replacements included the output board, one low frequency speaker driver, and a ten pound power transformer. The technician's work was good, but he was used to working on small amplifiers. The event warns us that high power amplifiers are in a class of their own when it comes to service, and even though the bias and balancing setup procedures may be straightforward, they need an experienced eye on the system as a whole.

The same thing applies to testing ragged looking equipment. Rather than risk a full power switch-on when checking an amplifier that has obviously been carelessly handled, it is wise to power up first with a 100 watt light bulb in series with the mains. If the initial flash due to normal **in-rush** current is followed by a dull glow, and if it remains stable when tapped, shaken, prodded, and run at low level, then it's relatively safe to power it directly for the full output test, given the cautions about some types of equipment detailed in Chapter 11.

Some switch-mode amplifiers are not safe to test this way because the series lamp won't prevent short circuits that will rapidly discharge the already full power supply capacitors. When this happens, large currents can pass through components that may explode with force. Before working on large power supplies that have been switched on within the last day, it is advisable to discharge the capacitors through a resistor. A small soldering iron element can be used in the absence of a 200 ohm 10 watt resistor, but it will take longer, and should be monitored with a voltmeter.

DUMMY LOADS

One of the most useful tools in amplifier work is a **Dummy Load** (Fig. 3.9). To make a stereo pair, follow the circuit in Fig. 3.8a, using two load resistors of 100 watt rating. The resistors get hot, and this will limit the period of the test, as their resistance will rise and invalidate the measurement. If you intend to measure very large amplifiers, the resistors can be mounted in an oil filled container to drain the heat. If used dry, a resistor encapsulated in ceramic compound will have high thermal inertia. Otherwise, mount the pair of 8 ohm resistors on a piece of half inch plywood with stand-offs to isolate the heat, together with small speakers on PCB screw-mounts connected via 1,000 ohm 5 watt series resistors.

Fit terminals or test sockets to enable you to clip your AC voltmeter across the load resistors when making measurements. A pair of flying leads with or without banana plugs or suitable amplifier output plugs completes the test kit, which is best made as a stereo or two channel dummy load because most amplifiers specify their output with both channels driven. The meter should be able to read AC volts RMS with accuracy. If in doubt, compare it with a reliable standard in the 50 and 250 volt AC ranges.

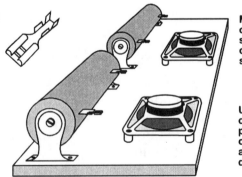

Quick-connect receptacles soldered on to each resistor terminal make sockets for meter and oscilloscope probes.

The resistors get hot! They need to be mounted off the board.

Mount a pair of four inch speakers on half inch stand-offs

Use a piece of half inch plywood or chip-board as a base for the dummy load.

Fig. 3.9 Stereo dummy load enables both channels to be fully driven. Worth many times its low cost.

A cable and connector for the oscilloscope input completes the test bay, and to ensure that the amplifier output cannot be accidentally grounded, include an isolating capacitor of 0.05 microfarads in each side of the oscilloscope line, unless the oscilloscope has a balanced input. Most amplifiers will have one side on ground, in which case there is only polarity to worry about, but bridge configuration output stages and bridge mono connected amplifiers have an output above ground on both sides.

CASCADING POWER AMPLIFIERS

There are many occasions when it is convenient to feed a power amplifier input from another amplifier's output; whether or not that amplifier is driving a speaker. Example: An amplifier doesn't have the output capability required, like a 30 watt center channel for a home theater system, when an 80 or 100 watt limit would sound cleaner.

A line output could be used, but perhaps it is not brought out to the rear panel, or is otherwise inaccessible. The answer is to connect an outboard 100 watt amplifier in cascade, from the speaker output. Many people have said to me, "You can't do that. Its too much power to go into an amplifier input."

Well, it's not a case of too many watts, because we know from the power formula that a 30 watt amplifier can only deliver about 15 volts across an eight ohm load, and across 20,000 ohms that's less than 12 milliwatts, or around a hundredth of a watt. The output voltage, however, might be the problem as it could be enough to overload the input stage of the following amplifier.

Of course, we could turn down the gain and feed only one volt of signal to the 100 watt amplifier, but the noise floor of the first amplifier would then be audible; a combination of white noise (hiss), hum, and buzz at mains frequency plus its harmonics. The first amplifier must operate at a nominal output of one or two watts to keep the noise floor in proportion, and gain reduction needs to take place after it, but before the input stage of the following amplifier is overloaded.

Broadcasters and telephone companies have made an art-form of amplifier cascading in studio systems and long line repeaters, they do it all the time, and we can do it too, by correctly placed terminations and attenuators.

All analog sound equipment from a transistor stage to a broadcast transmitter is subject to the immutable law that there's only about 80 dB

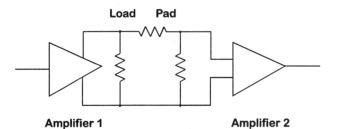

Amplifier 1   **Amplifier 2**

Fig. 3.10 **A pair of amplifiers cascade connected with output termination on the first and level adjustment for the second input.**

between system ground noise and peak level distortion. So it is important to present any equipment with program at a level that falls ideally between its usable volume limits. Audio modules in a system have to be driven at line level where it is appropriate, not forgetting that hi-fi and professional line levels are different.

To keep the noise floor at the correct distance from the signal, a loss pad (fixed attenuator) is inserted between the amplifiers. The source amplifier also needs a termination, to ensure stable operation within its specification. Figure 3.10 shows the circuit.

A fuller explanation of noise thresholds is found in Chapter 4. Page 87 shows 200-ohm pad values to suit the following input circuit.

LINE AMPLIFIERS

While Power Amplifiers are concerned with driving loads like speakers, requiring several watts, there is another class of amplifier devoted to line driving. **Line amplifiers,** which are specialised power amplifiers usually of less than one watt output, deliver signal voltage from a low impedance source to promote noise immunity. Noise immunity is dealt with on page 48 under "Audio Lines" and in Chapter 10.

Line amplifiers are the output end of most audio processors, but they are also used as loss recovery modules and line drivers in long lines terminated in 600 ohms at power levels usually up to +24 dBm (a quarter of a watt). The low source impedance of an audio **transmission line** and the relatively high signal current through the cable and the load at the far end combats frequency response degradation due to cable capacity and inductance. Domestic line outputs generally suit terminations of 10,000 ohms.

Small power amplifiers can be used as line or distribution amplifiers simply by terminating them and padding the output down to a suitable voltage, as described in the section on Amplifier Cascading, with the proviso that the pad presents a suitably low impedance to the line as previously discussed.

A **distribution amplifier** is one which drives many lines, having a splitter network at the output, like a mixing network in reverse, to offer the outgoing lines some degree of mutual isolation. In individual equipment modules, a line amplifier's purpose is partly to act as a buffer, so that conditions encountered at the line and its load can not feed back into preceding stages. Distribution amplifiers may also have a buffer stage on each output.

The line driver principle can be extended to those occasions when existing unshielded cables have to carry line level signals. A small power amplifier drives the line at one or two watts into a 10 ohm terminating resistor at the far end. It's an extra cost, but there may be no alternative if existing unshielded speaker lines are available, and the building is finished to the point that no more cables can be laid.

Page 63 describes use of a battery powered microphone header amplifier to overcome line noise, and on page 94 the same system, powered from the main power supply, is described increasing the output from 70 mm film magnetic replay heads, which are located several feet from their preamplifiers in the presence of strong magnetic hum fields.

LINES ARE DRIVEN FROM A LOW SOURCE IMPEDANCE AND TERMINATED AT THE FAR END.

VOLTAGE AMPLIFIERS

Mention of voltage amplifiers often arises. All amplifying systems drive some kind of load, and output voltage doesn't often exist by itself, but voltage amplification is a term of convenience to describe, for example, a gain stage that exists within a complete amplifier, preamplifier, or mixer, to bring up a small signal or recover a loss. A line amplifier used for insertion loss recovery following a passive equalizer could be called a voltage amplifier.

A voltage amplifier's gain can be described in decibels provided the termination impedances before and after it are the same. But "voltage amplifier" often refers to a gain device with a high impedance output that is not intended to actually do any "work," in the sense of driving a load.

There are many loose terms in audio, and this one in particular will come into focus when the difference between **voltage gain** and **power gain** is discussed in Chapter 4.

AUDIO LINES

Cable for line level audio frequency has a shield to reduce induction of noise voltages and currents into the line. The shield also works to prevent cross-talk from adjacent lines, so stereo left and right channel signals never share the same shield, and in twin shielded applications both channels have their own cable. Installation measures that make the shield work to trap noise are detailed later in Chapter 10, but we're primarily concerned here with line level and impedance, and how to use it to reject ordinary degrees of noise interference from the power mains, and cross-talk from other lines.

Small noise currents enter a line directly by induction as audio frequencies, but also as RF radiated by power wiring. The wiring in a building is an antenna broadcasting the RF content of noise sources, and when this gets into amplifying stages, any non-linearities detect the RF and resolve it as audio frequency.

Since induced noise currents are comparatively small, they can be countered by using a line impedance low enough to short-circuit them. There is always *some* noise, but if it is kept in suitable proportion to the signal, then it is inaudible, and that is the purpose of driving the line with an amplifier of low impedance, and then terminating the line with a load to make signal currents high relative to the induced noise currents.

The lower the load resistance, the higher the current for a given line voltage. Typical line impedances are: 600 ohms for long lines, and 10,000 ohms for short runs; up to 50 feet, or 15 meters. Noise is best attenuated at source, but the first and last lines of defence are to drive lines from low impedance sources, and to make sure line levels are high enough to keep signals above the noise floor.

Low source and termination impedances also reduce high frequency loss in cables because the line impedance is then low compared to its shunt capacity (Fig. 1.2).

COAXIAL TRANSMISSION LINES

Two conductor overall shielded cable, like microphone cable, is used almost exclusively for line level audio, but coaxial cable, its purpose and how it works, is mentioned here because its function is more complex.

Ordinary shielded audio cable, whether it is single conductor (Fig. 3.11a) for short runs up to twenty feet, or two conductor cable (Fig. 3.11b) serving each channel for long runs and higher noise immunity, has moderate loss at normal impedances and frequencies.

A long cable run will attenuate high frequencies more than others due to the capacity between conductors and from conductors to shield, plus series inductance, but in a long line installation this is taken care of at the repeater amplifiers with a loss correction network.

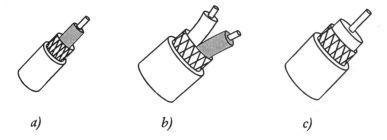

a)   *b)*   *c)*

Fig. 3.11 a) Single shielded cable. *b)* Twin shielded cable. *c)* Coaxial cable.

Coaxial cable (Fig. 3.11c) is quite different, and while short lengths of it can be used successfully for short run high impedance circuits due to its lower capacity without any specific termination, its main use is for cable transmission of RF (radio frequencies), video signals, and computer networking, which are all high frequency applications where frequency response and predictable signal delays are important. It works like this:

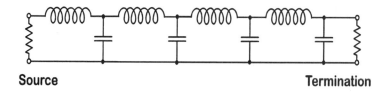

Source   **Termination**

Fig. 3.12 Circuit equivalent of a coaxial transmission line.

The series inductance and shunt capacity of the cable is designed to work with critical cable terminations, for example: 50 ohms for some computer networks and 75 ohms for receiver antennas and video signal cables. Correctly terminated, the length and shield spacing form an effectively infinite number of capacitance shunts and series inductors. These work as a multi-section **all pass filter** network to transfer a substantially flat response within a wide band of frequencies.

A coaxial transmission line, however, is not loss free. The loss is relatively high compared to an ordinary shielded cable that works independently of a precise termination for correct frequency response, because each theoretical filter section has its own insertion loss. There is another good/bad trade-off too, in the form of signal delay. Any series of coupled filter sections will act as a delay line.

Two video signals, or even a stereo sound channel, will suffer phase differences if they arrive through different length paths through any cable, but much more so through a coaxial transmission line. This effect is sometimes put to use where a small delay is required to correct an error between two lines. The signal in a transmission line can also be sampled at any point, or series of points, along its length, to distribute signals degraded only by loss of level, provided the line is correctly terminated at both ends.

Terminology is important where cable identification is concerned. The drawing of Fig. 3.11 will help to identify the difference, because one cable will not work properly in place of another.

Video Distribution Amplifiers and Data Networks invariably use a coaxial transmission line, and supply enough initial drive to counter the anticipated cable loss. The end of line terminator is the vital feature of this system, enabling input/output taps to be made at intervals along the line. If a coaxial network line appears to be functioning incorrectly, the first thing to check is the presence or otherwise of the end-of-line terminator, usually a resistor built in to a blanked-off connector.

HEARING AID INDUCTION LOOPS

In the 1950s, when a hearing aid was the size of a small notebook and was carried in a shirt pocket, the telephone pick-up coils were large and fairly sensitive. A 20 watt amplifier with the two ohm output connected to two turns of wire run around the perimeter of an auditorium or television stage would provide a sufficiently strong audio induction field for communication purposes. The miniature hearing aid induction coils currently used require four times the field energy to give satisfactory noise free results, as the perception of acceptable noise level has also changed.

Between 80 and 300 watts is used to drive induction loops, which range from a small loop under three or four rows of theater seats, to a loop surrounding an auditorium, or installed under the floor. Under-carpet cables are at hazard from carpet laying tools, and need some sort of protection, since the possibility of finding the break is remote.

Wall installations can be at floor level, but work best at four feet; the height of a seated person's head. A compromise system for theaters is to confine the loop to a small area where hard of hearing patrons can be seated.

An amplifier for loop driving applications needs to handle loads of one ohm, and many professional amplifiers will do this. One or at the most two turns are required for the loop, as multiple turn loops which would improve field strength also occasion high frequency loss.

The same procedure that was outlined on pages 35 and 36 of this chapter regarding attenuation of subsonics when driving constant volt line transformers, also applies to induction loops, but not for exactly the same reasons. A loop covering a whole auditorium will have enough resistance to make a practical amplifier load, but the nature of induction pick-up calls for attenuation at power mains frequencies, so it makes good sense not to waste power warming up the loop with frequencies that won't be heard.

DC COUPLED AMPLIFIERS

Many power amplifiers of respectable size have DC coupling from the output stage to the speaker line, but have a blocked DC path at the input.

While the power circuit at the speaker works much better without intervening capacitors or transformers, the input circuit is not handling power (to any extent), and capacitors are often inserted. This saves the system from various ills, eliminates the problem of DC offset voltages from previous equipment, and generally protects the amplifier and its load. There are, however, good reasons for DC coupling a system throughout, provided sensible precautions are taken to safeguard it, like mute-delay power-up. DC servo-amplifiers that drive electric motors have to output DC in order to function at near zero speed. There are also AC servo systems, but that's another story.

In audio, the advantage of optimum transient response, low phase error, and a bass response down to zero cycles per second, attracts many users, and there are systems which directly couple the preamplifier to the power amplifier to achieve these results. This is all well and good, but it does not mean that we can couple just any source to a DC amplifier.

The first indication of trouble may be that the speaker polarizes hard one way, and starts to burn out. A small DC input has caused a large DC voltage swing at the output, limited only by the supply-rail voltage, and several amps of direct current are now flowing through the voice coil. While servo amplifiers have current limiting to protect motors, this is not done with audio amplifiers. The victim will be the expensive bass driver, because mid and high frequency drivers are protected by the series capacitors in the crossover network.

The solution, if using non-specific or doubtful equipment to drive a DC amplifier, is to couple each channel with a polyester capacitor large enough not to limit the useful bass range. A side effect of this will probably be a loud bang as the capacitor charges when the system is switched on, but an effective muting or sequence circuit, which is found in most quality amplifiers, will keep the system quiet until all the glitches that accompany switch-on have passed.

As well as in DC amplifiers, muting and sequential switch-on are good precautions also where there are several power amplifiers in a system. This is because simultaneous power-up of several amplifiers and other pieces of equipment can produce fuse-blowing surges as a cascade of glitches plays havoc before feedback loops have woken up and pulled the various gains back to normal. In the absence of automatic muting, it is necessary to advise operators to power up amplifiers one at a time, in large installations.

Chapter 4.

Sound Waves and Decibels

THE AUDIO AND ELECTROMAGNETIC SPECTRA

Sound is wave motion in an elastic medium, a train of compression and decompression waves that travel in air at 1100 feet or 335 meters per second, at sea level, on a fine day.

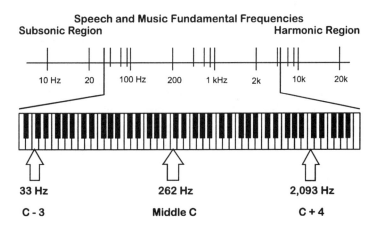

Fig. 4.1 The audio frequency spectrum.

As with all wave motion, no matter actually travels, just the effect, as one portion of the medium affects the next, which affects the next and so on. A speaker diaphragm makes sound waves by moving forwards and back at the audio rate. Several different sounds can be reproduced at the same time by a single diaphragm because many frequencies mixed together add and subtract at every instant to form a single complex wave.

This statement can be tested if you draw a graph of high and low frequency sine waves. Mark the rise and fall into sixteen or so equal vertical divisions on the Y axis; plus eight and minus eight, and sum the figures representing the two frequencies at about twenty equally spaced points on the horizontal X axis. When a third graph is drawn from the resultant figures, a complex wave will take shape. The same concept can be employed to visualize phase relationships and heterodyne frequencies; sum and difference frequencies generated by mixing two different frequencies. The waveform will contain elements of both frequencies.

Hertz is the term for **cycles per second**, and has been in use since the the establishment of the digital, or pulse technology era in the second half of the 20th Century, because audio, video, data, and radio signals are no longer just cyclic, but can be represented by sine wave, complex wave, or a data-stream of pulses. Hertz therefore describes a signal's repetition rate regardless of type.

The perceptible audible spectrum (Fig. 4.1) effectively extends from just under 20 Hz to 20,000 Hz. The higher frequency band is the domain of **harmonics**, mathematical relations of fundamental frequencies. Second harmonic is twice, or one octave above, the fundamental frequency; third harmonic three times, and so on.

Below and above the audible frequency range, sound is classed as subsonic and ultrasonic. Ultrasonics are a commercially and scientifically useful part of the sound spectrum, but the Audio Engineer's job is often to get rid of inaudible sounds before they cause trouble (filling up available recording space or amplifier power, and sometimes heterodyning with recording bias and sampling frequencies to form audible tones). The practical sound spectrum does not extend much beyond a few hundred thousand Hertz. Similarly, subsonics are often a nuisance, except where they play a part in movie soundtrack effects and ambience. Filters for dealing with various parts of the spectrum are described in Chapters 5 and 8.

THE 630 HZ TOOL

630 Hz is the exponential center of the audible spectrum, and has long been a useful design tool in consumer and commercial products. It's a key number; frequency of a car horn; power center of the human voice; dimensional center of the logarithmic chart in Fig. 4.1. Since most sound systems have end frequency limitations, especially the bass, it is very desirable to intentionally limit the other end accordingly. Note that the −3 dB point is usually quoted as the cut-off frequency in a frequency response specification, but the *effective* cut-off frequency is the one to

which we make reference here, as it takes into account other frequency characteristics like peaks and troughs throughout the response curve. The importance of shaping a response curve when designing a product is based ultimately on **how it sounds** on a variety of programs, and in comparison to other products and standards.

The 630 Hz spectrum center tool works like this: 630 squared equals approximately 400,000. Therefore a balanced spectral response will have end frequencies whose product is 400,000. The following classic frequency ranges have end frequencies that multiply together to equal approximately 400,000:

Wide range reproduction	20	to	20,000	Hz
Medium fidelity system	33	to	12,000	Hz
Pocket radio	153	to	2,600	Hz
Communication systems	333	to	1,200	Hz

A portable radio with a response from 150 to 20,000 Hz will sound unnaturally bright. with out of proportion upper harmonics. Harmonics are also the product of distorted fundamental frequencies, so it may also sound distorted, or even noisy. An AM broadcast receiver with a highly selective, narrow band receiver will possibly have little response above 3 kHz. If the bass is not similarly restricted it will sound bassy and unnaturally dull, although otherwise the response may be perfectly flat. Note that a high frequency peak within the **pass-band** may restore balance because it changes the **apparent cut-off frequency**. The 630 squared tool is also very useful when mixing dialog and effects. Chapter 8, on mixing and **signal processing,** has more information on this subject.

Sound travels in solid, liquid, or gas. Its traveling speed and efficiency vary with the medium and its range is limited by the scattering and damping effect of the medium in which it travels.

The **Electromagnetic Spectrum**, shown in Fig. 4.2, is different. It propagates its waves equally well in a vacuum, but is reflected by almost anything in the way. EM wave motion moves at the speed of light, part of the same spectrum. It loses intensity by the square of the distance if it radiates freely, but can be directed and concentrated into a beam in which the inverse square law is not relevant.

Because the high frequencies of the EM spectrum travel fast and can span great distances, they are used as the medium for radio transmission, where a coherent, or single **radio frequency** carrier wave is modulated either in amplitude (AM), frequency, (FM), phase, or pulse code (PCM),

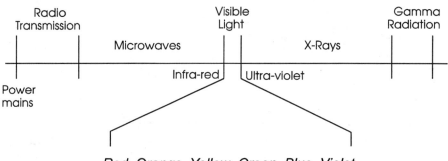

Fig. 4.2 **The electromagnetic spectrum.**

at the rate of the audio signal. Whereas the audio spectrum is limited to a few thousand Hertz, the EM spectrum, or spectrum of waves propagated by an **alternating magnetic field**, commences from zero; a stationary magnetic field, to the top end of the radio spectrum thousands of megahertz distant; beyond that through Infra-Red emission, then the visible spectrum of Red, Orange, Yellow, Green, Blue and Violet light (which colors sum to make White Light), to Ultra-violet and beyond. Although its effect and the distance it can travel varies with frequency, it is all the same EM spectrum as the familiar medium of radio transmission. Masers and lasers respectively belong to the microwave and light bands.

The extent of the Electromagnetic Spectrum is shown by the wavelengths of the various bands; a more useful comparison than frequency because wavelength statements avoid the very large quantities that would have to be used to describe the frequencies involved. The wavelength of a frequency is the distance traveled during one cycle:

Radio Communication	30,000 meters to 0.1 mm
Infra-red Light	0.4 mm to 0.0007 mm

We have to switch to smaller units here, as the decimal places would get out of hand: 0.0007 millimeters equals 700 nanometers, or 7000 angstrom units.

Visible Light	7000	to	4000	angstroms
Ultra-violet Light	4000	to	120	angstroms
X-rays	120	to	0.06	angstroms
Gamma Radiation	1.5	to	0.01	angstroms

The Angstrom is a convenient unit used to describe the frequency of the colors of visible light, among other things. The angstrom dimension can be imagined if **Newton's Rings** are demonstrated: visible rainbow rings that join areas of equal distance when two sheets of transparency film are placed together. The colors are interference patterns produced at the wavelength distance for each color, between the transparency sheets.

FREQUENCY AND WAVELENGTH

In both the audio and radio frequency bands, there are two main methods of addressing a position in the spectrum: **frequency** and **wavelength**. Both are useful. Frequency describes the pitch of a signal in Hz or cycles per second, and wavelength describes the dimensional length of each cycle, or the distance between identical parts of adjacent cycles, which is the same thing. Audio wavelength is usually stated as being in air at sea level; earth normal pressure and temperature, in any convenient unit of length.

RF (radio frequency) wavelength is given in meters, or in centimeters for very high frequencies. The numerical values of wavelength and frequency are inverse; as frequency increases, wavelength becomes less. The normal wave travelling speed is 1,100 feet or 335 meters per second for **sound**, and 86,000 miles, or 300,000,000 meters per second for **EM** waves. In audio and RF design work it is often necessary to know the wavelength of a particular frequency, and these wave speeds enable us to convert very simply from frequency to wavelength and vice versa, using the following formulae:

Audio: $Hz = \dfrac{335}{meters}$ $\qquad Meters = \dfrac{335}{Hz}$

RF: $Hz = \dfrac{300,000,000}{meters}$ $\qquad Meters = \dfrac{300,000,000}{Hz}$

PROPAGATION

Both radio waves and sound waves travel in a straight line in their respective media if there's nothing in the way, but they can both be reflected and refracted. Otherwise, radio and sound from omnidirectional sources obey the inverse square law which dictates that intensity falls as the square of the distance, all other things being equal. **Reflection** means bouncing off a body, and like light off a mirror, both radio and sound waves leave the reflecting surface at the same incident angle as they approached it. **Refraction** means that the direction of the wave's travel is changed, or the rays are bent. This happens, just as it does in light, by interference of suitable conditions. Radio waves are bent and redirected by magnetic fields and wave-guides. Both light and sound rays are bent when they pass at a suitable angle through changes in density of the medium they pass through.

Transparent lenses refract light; changing air densities, or the presence of solid materials, change the propagation angle of sound waves. Some speakers use horns to amplify and redirect the sound. Others use acoustic lenses which produce similar dispersion properties by shaping the radiation angle.

Radio waves, light, and sound waves all radiate in a pattern as the source and its attachments direct. Broadcasters shape the output of their transmitter antennas to avoid areas of low population density and open sea. However, they concentrate their output in the direction of cities, freeways, and other high listener population areas.

Sound reproducers, or speakers, also direct their output where it will do most good, since the available energy that they radiate is not unlimited, and it can also be a nuisance if too much of it goes in the wrong direction (as will be seen in Fig. 6.8).

The propagation angle of a speaker should ideally be the same at all frequencies, and that is one of the aims of the designer, but at low frequencies sound is substantially omnidirectional, and does not obey exactly the same laws as the middle and upper frequencies. There are a variety of maneuvers that speaker installers can employ to solve the problem of bass dissipating itself in wasteful and undesirable directions, and these will be discussed at length on pages 111 and 112.

AUDIO MEASUREMENTS

The main parameters of audio electronic equipment are: **Gain, Frequency response, Distortion, Noise,** and **Power output**. Given any piece of electronic transmission equipment, it is possible to readily discover it's performance in these five areas without knowing anything else about it, other than the power supply requirement. Power output has been dealt with separately in Chapter 3, because it is the most obvious maintenance activity where power amplifiers are concerned. The other parameters can also be determined by sending an audio signal or series of signals into the input, and measuring the resultant output.

Unless it is a preamplifier or power amplifier, whose purpose is to change levels, it can be assumed that many audio modules including recorders are designed to have unity gain. There are exceptions, like equalizers, limiters, and filters, but even in these non linear devices the principle of unity gain is intended to apply to the spectrum or level range unaffected by them. Line amplifiers are another exception, used for two main purposes: as **line drivers**, and as **recovery amplifiers** to accommodate losses encountered at faders, passive equalizers, and so on. Line amplifiers usually have gain variable from zero to between +6 and +35 dB. A loss can be a device's **insertion loss**, a fader's "holding" loss, which, once recovered, is therefore its **available gain**, or it may be the loss attributable to a resistive mixer or splitter network.

Statement of unity gain pre-supposes normal line levels, and this means that the signal **voltage level** at the output will be the same as at the input, given central, or "reference" set-up of any controls. Voltage level, because electronic modules normally have bridge inputs and power outputs that are capable of driving loads between 10,000 ohms and 600 ohms, the most common line impedances.

A good example of unity gain would be a mixer, where a line input of 1.228 volts (+4 dBm in 600 ohms) at a given frequency would produce an output of 1.228 volts **at the same impedance** provided the channel input gain preset controls were set to **line**, the channel and master faders were set to "Hold" 10 dB, or **zero**, and all equalizers were set to the **flat** frequency response position. Holding between 6 and 10 dB is a normal fader condition, which means that the operator has the option of increasing gain by up to that amount or reducing gain by 120 dB. Having established gain or loss at a central frequency, we are then in a position to compare the gain at other frequencies; that is, perform a frequency response check, measure the distortion at both line level and peak level of +10 dB; and measure the system noise below peak level, by cutting the source after establishing a peak level tone. The input should remain terminated to validate the measurement.

Specifications (Fig. 4.3) are as exacting as they can reasonably be maintained in recording or multi-generation work, because each frequency response error, distortion, or noise floor adds to the next one, as can be seen from quality loss in tape copies. Where noise or distortion figures add, the square root of the total noise is equal to the sum of the squares of the individual figures in dB, by the equation:

$$Total\ Noise,\ dB = \sqrt{a^2 + b^2 + c^2}$$

59

	Monitor amplifiers	Mixers and line amplifiers	Recording equipment
Frequency Response			
	+/- 1 dB	+/- 0.5 dB	+/- 0.5 dB
Total Harmonic Distortion at Peak Level			
	1.0 %	0.1 %	3.0%
Noise below Peak Level			
	50 dB	80 dB	65 dB

Fig. 4.3 An example of audio performance minimum specifications.

NOISE AND DISTORTION THRESHOLDS

There is always noise present in any electronic analog system; even in a piece of wire a low noise level is generated by random electron movement at any temperature above absolute zero. The noise floor, or ground noise, of an audio module is made as low as possible, consistent with other features of the design, since many choices in electronics are an educated compromise between two or more factors, including cost effectiveness in the equipment's intended market.

There is also a ceiling level under which the program must fit. So a program of a given dynamic range has to operate centrally between the limits of noise and peak level distortion; two of the main limitations of any system. The third limitation is bandwidth, or effective frequency range of the system, as this can be an occasion for distortion if high frequency pre-emphasis is employed, or where inclusion of a speaker processor, controller, or equalizer, decreases the headroom of the system at the boosted frequencies. Level metering of recording systems often includes a spectrum monitor with a line drawn indicating the maximum safe level at every frequency like the examples that will be shown in Figs. 5.4a or 6.19.

Fig. 4.4 shows the dynamic range of the program positioned between the limits of noise and distortion; a concept that might seem too obvious to mention, until the situation arises that a group of two or more processing modules; line amplifiers, mixers, equalizers, or whatever; are assembled, and one of them is suspected of being **noisy** (or **distorted**; see Case Histories, pages 68–69).

Every piece of audio equipment on the market has been through a careful design procedure to balance levels, gains and losses at every point

Fig. 4.4 Relationship of program and system dynamic ranges.

along the way from input to output, including DC paths, negative feedback loops, and choice of input and output devices, to ensure that performance will meet specification under all conditions.

The sort of contingencies the designer anticipates would include component aging, high or low ambient temperatures, unusual input and output terminations like highly inductive or capacitive loads. Further, the manufacturer intends the equipment to perform correctly at the chosen frequencies, and to be unconditionally stable up to ten times the pass-band. All this careful design and specification, which all adds to the cost of the unit, is wasted if an audio module is included in an incorrectly level-aligned chain, or an insert point at the wrong level.

Ninety five percent of the time there's nothing wrong with any of the units. If noise is the problem, one of them is probably being driven at too low a level, and its normal ground noise will therefore be audible. Referring to Fig. 4.4 again, the program dynamic range has moved to the left at the part of the chain in question and its low level program components are buried in normal ground noise. This can not be extracted by turning up the gain later, because the noise will also rise. The answer to this, and all such problems involving noise or distortion, is to make sure every item in the chain is being driven at its specified input level. Refer to page 63, Decibels and Power, for the math which supports the following discussion.

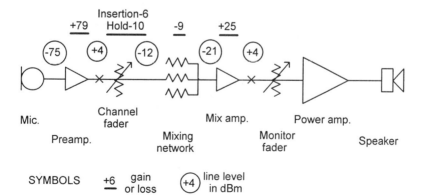

Fig. 4.5 Example of gains, losses, and levels in a mixer.

The gains and losses in dB (underlined), represented in Fig. 4.5, balance each other in order to maintain line levels in dBm (circled), wherever possible, not lower than minus 30 dBm. If levels are allowed to fall too low, there will be a noise penalty. Similarly, if line levels are permitted to rise too high, distortion may occur either at the output of a line amplifier or at the first stage of the following one. The operator should therefore be aware of line levels at the input and output of every module in the chain. In practice this is seldom a problem if every module is has unity gain, but testing is simple, and is a splendid reason for owning a signal generator and an AC millivoltmeter (Fig. 4.6). When applying test tone, normal input termination should be maintained.

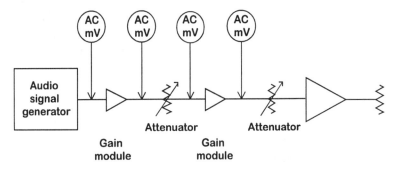

Fig. 4.6 **Measuring line levels in an audio chain.**

Commonly used line levels are 1.228 volts (+4 dBm), 1.55 volts (+6 dBm) in 600 ohms or (domestic products) half a volt in 10,000 ohms. Consistent with a +10 dB peak factor above program zero level, maximum output capability for line amplifiers varies between +18 and +24 dBm. Special line amplifiers for use as splitters or for anticipating passive device insertion loss may output up to +30 dBm (1 watt).

The peak factor mentioned is allowed for in some types of level meter due to the mechanism's mass and inertia, which causes the dynamic program meter reading to be lower than the true level. Among these, VU (volume units) meters, identified by a yellow scale, are made with defined ballistics to standardize the instruments with an achievable specification. Further information is given on page 160, where the meter's ballistics are discussed as a factor in reading the meter.

The central frequencies used for quick level checks vary from 400 Hz to 1 kHz. A quick frequency response check would be at 100 Hz, 1 kHz, and 10 kHz, sufficient to pick up gross errors. The thirty one frequencies shown in Fig. 6.19 are used for precise response calibration.

The foregoing comments on noise thresholds do not apply to microphone preamplifiers because they are designed specifically to work with low input levels. However, there are times when it is desirable to increase the level in a microphone line. First, there may be noise interference getting into a microphone cable from lighting and other power sources, or the microphone output may be quite low in drama production work. Second, a noise signal that is above or below audible frequency may get into a microphone channel and false-trigger an automatic attenuator system, whether it is used to gate the microphone on, or attenuate another program when the microphone is used. Third, it may be desired to use a high impedance microphone with a long line. In these circumstances, it is practical to fit a "header" amplifier at the microphone end to drive the line at a higher level (Fig. 4.7).

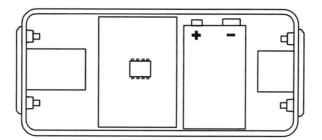

Fig. 4.7 Layout of a +10 dB microphone header amplifier.

A header amplifier consists of either two transistors or one IC mounted in a small diecast 'jiffy' box with a battery and connectors. Note that the battery needs to be shielded as well as the amplifier. The current drain is very low, and the battery will last several days at least, so if the production warrants it, fitting a fresh battery every week or every day is not a great expense. Voltage amplifier modules are available from electronic accessory suppliers. Further discussion on the use of small gain modules appears on page 94.

DECIBELS AND POWER

The loudness unit called the **Bel** was named after Alexander Graham Bell, inventor of the telephone. The Bel originally represented the power loss in one mile of standard telephone cable, and since the human ear's perception of loudness is logarithmic, the ratio between two sound pressure levels in Bels is the log of the power ratio. The **Decibel** is sometimes considered to be the smallest perceptible level change, a more appropriate unit for advanced audio technology, and is one tenth of a Bel.

The decibel is the basic unit of increase or decrease in sound power, or intensity, and its function is logarithmic because the ear's perception of changes in sound level is logarithmic. Our hearing sensitivity to loud sounds is lower than to quiet sounds, enabling us to appreciate the rustle of leaves, or recognize every instrument in a live band. Up to the useful limit of hearing, it allows us to make detailed analysis of low and high level sounds, yet it does not detract from our ability to judge perspective.

Gain and loss expressed in decibels are **power** functions calculated from the before-and-after voltages measured across line terminations (see Amplifier Output Measurement, page 39). The **dBm power level** scale references all levels to 0.775 volts in 600 ohms (one milliwatt). The Decibel Equivalents Table on page 70 (Table 4.1) gives the ratio between the voltage at a given power level, and zero dBm, or 0.775 volts. The calculated power ratio is converted to decibels for convenience.

Power gain or loss expressed in decibels is therefore ten times the log of the power ratio:

$$dB = 10 \times \log \frac{W_1}{W_2}$$

Decibels state **the ratio between two audio power levels** whether it is a sound pressure level (SPL) change at a microphone or speaker, or power level gain or loss in an audio line. Increase or decrease of a power ratio by a multiplying or dividing factor is expressed by adding or subtracting decibels, because decibels are a logarithmic quantity. Referring to **dBm** column of Table 4.1, it can be seen, for example, that adding +3 dB to a level of +25 dBm to make +28 dBm multiplies the power approximately by a factor of two; from 0.316 to 0.631 watts. Doubling the power adds 3 dB. Halving the power subtracts 3 dB.

Another example will help to clarify the concept of the logarithmic decibel scale. Again, in the decibel table, it will be seen that at the high power end of the scale, an increase of 3 dB makes the difference between 50 watts and 100 watts. Twice the power is still a three decibel increase. It is apparent then, that +3 dB can represent a 50 watt increase at the same time as it represents an increase of 0.315 watts, because an added logarithm is a multiplier.

This demonstrates the compactness of logarithmic representation. The logarithmic audio frequency chart at the beginning of this chapter effectively displays 10 Hz to 20,000 Hz in the space of a few inches. In the same way, the dB scale converts a very wide range of power levels into an accurate and easily handled numeric system.

Decibels can be conveniently added or subtracted because they are logarithmic, otherwise, power level comparisons would have to be multiplied or divided. The other essential fact is that decibels follow the logarithmic sensitivity curve of our hearing ability.

DECIBELS ARE POWER DIFFERENCES

dBm is a power-level term used only for 600 ohm circuits. Table 4.1 compares levels from –20 to +50 dBm with related quantities and ratios, referenced to one milliwatt in 600 ohms, which is zero dBm.

DECIBELS FURTHER EXPLAINED

Table 4.1 at the end of this chapter displays a wide range of levels from 77 millivolts to 245 volts across a 600 ohm termination, representing power levels from a hundred thousandth of a watt to 100 watts.

These are real watts: 100 watts is enough to illuminate the average room if it's powering a lamp, or fill a theater with sound if it's driving a speaker. Watts of energy are the same, no matter what their use. Despite the enormous difference in the above power levels, the entire range is accurately addressed by a scale of just 70 decibels.

Column one, **dBm**, is the power level in decibels relative to zero dBm, the **reference level**, but not necessarily standard line level; that's more likely to be +4 or +6 dBm. The term dBm is always referred to 600 ohms, a standard transmission line impedance, and the zero dBm reference of one milliwatt in 600 ohms can be verified using the power formula from Chapter 3:

$$W = \frac{V^2}{R}$$

Zero dBm is:

$$\frac{0.775 \times 0.775}{600} = 0.001 \quad \text{watts}$$

Note that expression of a **power level** in decibels needs a zero dB reference to validate it. For example, the term **dBV** has in the past been used to express decibels referred to one volt in high impedance lines.

POWER GAIN OR LOSS EXPRESSED IN DECIBELS IS TEN TIMES THE LOG OF THE POWER RATIO.

To measure gain or loss and express it in decibels, it would seem necessary to go through several steps:

- Measure the before-and-after voltages.
- Convert them to power.
- Calculate the power ratio.
- Get out the log tables and convert the power ratio to decibels.

But there is a more convenient, and just as valid, way to measure and express level changes, or differences, and that is as **Voltage Gain or Loss.** From page 64, power gain in decibels is ten times the log of the power ratio. The short-cut method is to calculate dB directly from the voltage ratio:

VOLTAGE GAIN OR LOSS EXPRESSED IN DECIBELS IS TWENTY TIMES THE LOG OF THE VOLTAGE RATIO.

Ten times? Twenty times? Is voltage gain then twice the numeric value of power gain?

No it is not, it's just that the same result is obtained in alternate ways using the **different quantities** of power and voltage. The voltage ratio column in Table 4.1 compares a particular voltage level with the zero dBm reference, and the ratio is derived from that voltage divided by 0.775. The two voltages can be converted to watts with the formula:

$$W = \frac{V^2}{R}$$

However, it is practical to calculate power differences in decibels by voltage alone, because in this case, reference to load impedance is not necessary **provided it remains the same.** The practice has given rise to the term **"voltage gain,"** where instead of squaring the voltages as in the power formula:

$$dB = 10 \times \log \frac{V_1^2}{V_2^2}$$

it is easier to double the logarithm multiplier instead. The expression produces the same result and becomes:

$$dB = 20 \times \log \frac{V_1}{V_2}$$

Mathematical proof that the two methods produce exactly the same result is simple: Take two power levels with an approximate 2:1 ratio from Table 4.1, say, 50.11 watts and 100 watts. Note the 3 dB difference.

Calculate **10 times** the log of the ratio of the **square of the voltages**, and note that the result is the same as the calculation: **20 times** the log of the ratio of the **voltages**. The result is 3 dB either way.

Halve 100 watts and note that the loss is 3 dB.
Then halve the **voltage** relative to 100 watts, and note that the loss is 6 dB, but **power** falls four times to 25.11 watts.

Decibels are always decibels no matter how they are derived. The following facts are evident:

1. The two formulae for decibels given on page 66 produce the same result but use different quantities: power and voltage.
2. Multiplying a power level (watts) by two represents an increase of approximately +3 dB.
3. Multiplying a voltage level (volts) by two represents an increase of approximately +6 dB.

*600 OHM LINE POWER LEVELS ARE RELATIVE TO
ZERO LEVEL; ONE MILLIWATT IN 600 OHMS.*
Symbol: dBm

*POWER GAIN OR LOSS EXPRESSED IN DECIBELS
IS TEN TIMES THE LOG OF THE POWER RATIO.*
Symbol: dB

*VOLTAGE GAIN OR LOSS EXPRESSED IN DECIBELS
IS TWENTY TIMES THE LOG OF THE VOLTAGE RATIO.*
(Provided the circuit impedance remains the same)
Symbol: dB

CASE HISTORY

Shutting the door after the horse has bolted.

VENUE: A Dance Academy of 30 years standing.

PROBLEM: Low frequency distortion.

INVESTIGATION: Amplifier and speakers were independently tested.
                             No faults were found.

CUSTOMER LIAISON:

Discussions with the owner-operator revealed that all recommendations regarding speaker placement and type had been observed, and that the tone controls were appropriately set.

When the owner was asked if a new CD player had recently been purchased, the answer was "Yes, a top brand." It was then established that the main amplifier volume control was normally set at around 9 o'clock, and that the CD player had its own gain control.

It is a fairly common problem that some brands of CD player, especially used with discs that have substantial bass content, have a lot more output than earlier sources like LP vinyl records and players that people have been used to for a long time.

Explanations that the amplifier input stages were distorting before the volume control, and that the CD player should have its gain reduced, and the main amplifier gain control set higher, around 11 to 12 o'clock, failed to get through to the owner. Higher volume setting surely could not be the cure for distortion!

Finally, with the suggestion that it would be "no use shutting the stable door after the horse had bolted," the dance academy operator understood the need to set appropriate levels at all points along the equipment chain (Pages 60–62). Problem solved.

CD players without gain controls can be adjusted with a line level loss pad in the output circuit (Fig. 5.11). Main amplifier volume controls set very low on their scale indicate a risk of distortion at the input stage, before the control takes effect.

CASE HISTORY

Hazards of arbitrarily adding an insert point.

VENUE: A Powered Lectern, recently installed in a Public Hall.
PROBLEM: Audible hiss level.
INVESTIGATION:

The lectern, comprising a two part transportable cabinet with the amplifier and microphone preamplifier in the top half, and a single speaker in the bottom section, was taken to the service department for high noise level diagnosis and correction.

It was found that a graphic equalizer had been fitted at the owner's request to boost the speech presence region, since the unit operated on the floor instead of a raised stage, and the single speaker was positioned below the listeners. The lectern had no insert point for a signal processor, so the equalizer had been connected immediately before the power amplifier input.

The equalizer was found to be operating below normal line level, hence the audibility of its noise floor. To remedy this, the preset gain control on the equalizer was advanced, and its output correspondingly padded back until the hiss became inaudible for all practical purposes. The equalizer bypass switch was disabled to avoid a volume increase if it was accidentally used.

Many integrated amplifiers, both domestic and commercial, have no insert point, so if a limiter, equalizer, or acoustic feedback suppressor needs to be fitted, an insert point can be created in the form of Left and Right PREAMP OUT connectors, and Left and Right MAIN AMPLIFIER IN connectors.

However, the question of levels is important, as sufficient drive has to be applied to the following equipment. A pair of low gain buffer amplifiers should be added if necessary, to bring the preamplifier output up to line level, and the amplifier input padded accordingly.

Alternative terminology for insert point labeling, more suitable for studio mixers: INSERT SEND, and INSERT RETURN (left and right). Insert points are linked by jumpers or jack normals when not used (Fig. 5.16).

Table 4.1 Decibel equivalents.

dBm	Volts in 600 ohms	Voltage ratio	Watts
-20	0.077	0.1	0.00001
-19	0.087	0.112	0.000013
-18	0.098	0.126	0.000016
-17	0.109	0.141	0.00002
-16	0.123	0.159	0.000025
-15	0.138	0.178	0.000032
-14	0.155	0.199	0.00004
-13	0.173	0.224	0.00005
-12	0.195	0.251	0.00063
-11	0.218	0.282	0.000079
-10	0.245	0.316	0.0001
-9	0.275	0.355	0.00013
-8	0.308	0.398	0.00016
-7	0.347	0.447	0.0002
-6	0.388	0.501	0.00025
-5	0.436	0.562	0.00032
-4	0.489	0.631	0.0004
-3	0.548	0.708	0.0005
-2	0.615	0.794	0.00063
-1	0.69	0.891	0.00079

Table 4.1 Decibel equivalents *continued.*

dBm	Volts in 600 ohms	Voltage ratio	Watts
0	0.775	1:1	0.001 (1 mW)
+1	0.869	1.122	0.0013
+2	0.975	1.259	0.00123
+3	1.095	1.413	0.00195
+4	1.228	1.585	0.00251
+5	1.377	1.778	0.00316
+6	1.545	1.995	0.00396
+7	1.732	2.239	0.00501
+8	1.946	2.512	0.00631
+9	2.183	2.818	0.00794
+10	2.45	3.162	0.01
+11	2.748	3.548	0.013
+12	3.084	3.981	0.016
+13	3.474	4.467	0.02
+14	3.883	5.012	0.025
+15	4.356	5.623	0.032
+16	4.887	6.31	0.04
+17	5.484	7.079	0.05
+18	6.153	7.943	0.063
+19	6.904	8.913	0.079

(Table 4.1 continues on page 72)

Table 4.1 Decibel equivalents *continued.*

dBm	Volts in 600 ohms	Voltage ratio	Watts
+20	7.746	10:1	0.1
+21	8.691	11.22	0.126
+22	9.752	12.59	0.159
+23	10.949	14.13	0.22
+24	12.275	15.85	0.251
+25	13.773	17.78	0.316
+26	15.454	19.95	0.398
+27	17.232	22.39	0.501
+28	19.458	25.12	0.631
+29	21.83	28.18	0.794
+30	24.495	31.62	1.0 Watt
+31	27.49	35.48	1.259
+32	30.84	39.81	1.58
+33	34.74	44.67	2.0
+34	38.82	50.12	2.512
+35	43.56	56.23	3.162
+36	48.87	63.1	3.981
+37	54.84	70.79	5.012
+38	61.53	79.43	6.31
+39	69.04	89.13	7.943

Table 4.1 Decibel equivalents *continued.*

dBm	Volts in 600 ohms	Voltage ratio	Watts
+40	77.46	100:1	10 Watts
+41	86.91	112.2	12.58
+42	97.51	125.9	15.84
+43	109.49	141.3	19.95
+44	122.75	158.5	25.11
+45	137.73	177.8	31.62
+46	154.54	199.5	39.81
+47	173.23	223.9	50.11
+48	194.58	251.2	63.1
+49	218.3	281.8	79.43
+50	244.95	316.2	100 Watts

Chapter 5.

Transfer Characteristic and Networks

TRANSFER CHARACTERISTIC AND DISTORTION

When a signal is passed through an audio process, through an amplifier, speaker, line, or other; it can be modified contrary to the intention of the user by incorrect alignment or through a fault in the equipment.

The event that produces an output when the input is presented with a signal is called transfer, and the transfer characteristic determines the similarity of the two waveforms. It can be represented by a graph, as in Fig. 5.1, and seen as a line linking the two signals. The diagonal line can be readily displayed on an oscilloscope by bridging one signal with the X input and the other with the Y input, and adjusting the oscilloscope gains so that the line lies at 45 degrees. In this way the change in the line's shape can be seen as, for example, the signal passes in and out of clipping during amplifier testing, or other non linearity is revealed by curvature of the line.

The X-Y display needs more explanation. Most amplifiers and signal processing modules are designed to be phase coherent; that is, a positive signal peak at the

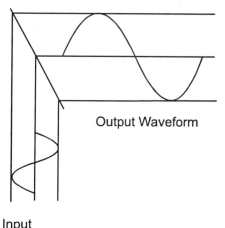

Output Waveform

Input Waveform

Fig. 5.1 Ideal transfer characterisic.

input will produce a positive peak at the output. In the X-Y test, zero and 180 degree phase relationships between input and output waveforms are indicated by a single straight line, but if identical signals from two tone generators can be made to drift apart in phase, the line starts to form a circle. A stereo signal pair on X-Y inputs of an oscilloscope makes a very informative display, showing differences in phase as well as levels. The X-Y inputs display many other functions. Pages 26–28 have more information.

The non-linear transfer curve in Fig 5.2 has reduced the comparative height of one half of the waveform. While it still has a rounded shape, it's no longer a simple sine wave; instead of one peak on one half-cycle, there are now two peaks, or rapid angle changes.

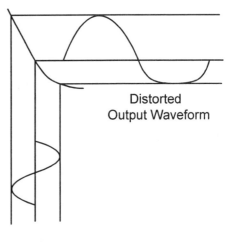

Distorted Output Waveform

This degree and type of waveform distortion depicted would result in unacceptable sound quality due to the addition of harmonics, the factors that individualize natural sounds. For example, the human voice and stringed instruments are rich in second harmonics, making relatively high distortion content tolerable in some cases; but flutes are very pure, and any distor-

Fig. 5.2 **Non-linear transfer characteristic.**

tion is obvious. Birds like canaries have a high third harmonic sound, an effect that otherwise seldom occurs in nature, so its addition to music or speech due to transfer or frequency response errors can be recognized as unnatural.

HARMONIC DISTORTION IS THE ADDITION OF HARMONICS

Harmonic distortion is only one type, and it is relatively easy to detect and measure in steady state tests as THD (total harmonic distortion), but there are more virulent forms like Transient Intermodulation Distortion, which occurs briefly and does not show up on steady state tests. TIM occurs, for example, when a program peak escapes from the output stage of an amplifier before the negative feedback loop has had time to transfer output level information back to the input to correct the gain. Amplifier stability can be upset by combinations of capacitive and inductive reactance in the output circuit.

Every stage in an electronic system introduces some delay. Nothing happens instantaneously, including the process of transfer. Some **MOSFET** and vacuum tube output stages are inherently low in TIM, but many amplifiers obtain their results using good design with bipolar transistors.

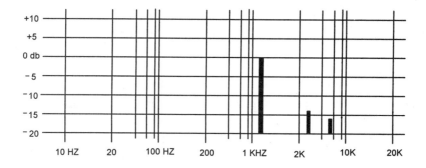

Fig. 5.3 Graph displaying a fundamental frequency and its second and third harmonic. A high resolution spectrum analyzer would be needed to observe single frequencies.

Fourier Analysis is a way of interpreting waveforms which defines their harmonic content. All sound can be said to be made up of little bits of sine wave. In the example suggested by the exponential spectrum graph of Fig. 5.3, it can be seen that a particular sound might be made up of sine waves belonging to the following frequencies; the 1.2 kHz fundamental; a lesser amount of 2.4 kHz second harmonic, and a still audible content of 3.6 kHz. Second harmonic is twice the fundamental frequency; third harmonic is three times the fundamental, and so on. Many harmonics are high enough to be beyond the usable spectrum, and for this reason, bandpass filters follow many signal processing events, especially of dialog, because it enables enhancement of those parts of the spectrum which promote intelligibility and naturalness, while attenuating harmonics that have been accentuated by the process, and which would otherwise work contrary to the production of clear, intelligible, and pleasing sound (page 169).

Undesirable harmonic levels can cause noise and distortion if they are outside the response capability of the recording or transmission medium.

When sound is recorded or reproduced out of normal perspective, like a person speaking in close-up, and especially when it is reproduced "larger than life" in an auditorium, with its enhancing reverberation and multiple reflections of transients, the harmonics can often appear to be out of proportion. The voice harmonics are good, an essential part of each voice's character, and the harmonics have to be preserved. But especially where compression is used, the upper harmonics often need modification to keep them in proportion.

Harmonic distortion is addition of harmonics; multiples of the fundamental, or original frequencies. So natural harmonics that fly away out of proportion to the rest of the program are going to sound very much like distortion; they sound bad, and mask intelligibility. The methods of dealing with this effect; making voices sound normal, is covered in Chapter 8, which deals in depth with signal processing, a major department of audio work.

Returning to transfer distortion, a wave analyzer details the frequencies present in the sample, and high resolution spectrum analyzers display them as graphs. For an Audio Spectrum Analyzer to do this, it will need to have a digitally processed display capable of a resolution of one cycle per second. The simpler spectrum analyzers display between ten and thirty one frequency bands with level markers, linked or separate, on an oscilloscope screen, or show them on an LED matrix (Fig. 5.4a). In most cases, the frequency bands of these analyzers have a high degree of sensitivity to adjacent bands, as shown in Fig 5.4b, and are suitable for displaying frequency response, but not for distortion measurement or harmonic analysis.

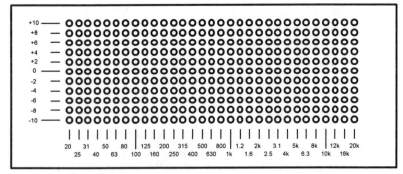

Fig. 5.4a LED (light emitting diode) matrix spectrum analyzer, which does not have sufficient resolution for distortion measurement.

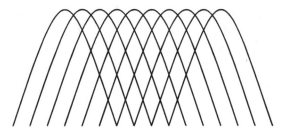

Fig. 5.4b Frequency band overlap of spectrum analyzer, similar to a graphic equalizer.

Small amounts of noise and distortion accumulate over several generations, or multiple passes through electronic modules, and the final result at the end of a boosted transmission line or a re-recording event of several generations will be good provided the individual modules are within a suitable specification. Multi-generation work requires higher specifications in individual audio modules than the final result would indicate, and is the reason for the exacting standards of professional audio products.

More common forms of distortion analysis are performed with a noise and distortion meter. It rejects, or nulls, the fundamental frequency in the tone sample and reads what's left as a percentage of the original (Fig 5.5).

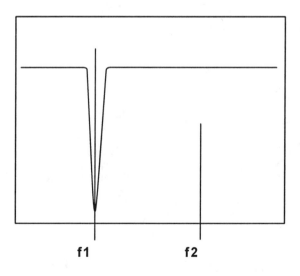

f1   **f2**

Fig. 5.5 Distortion meter nulls fundamental frequency, leaving harmonics plus noise.

THD read by this means contains noise as well, but is a practical way to speed measure a lot of equipment in a day. The **noise and distortion percentage** can be logged as a single figure, and detects gross errors and performance drifts in general, working on the theory that harmonic distortion is a common form of distortion, and indicates the presence of associated faults.

ELECTRO-MECHANICAL AND ACOUSTIC TRANSFER

In addition to the electrical transfer characteristic described on pages 75–76, audio signals are subject to two other transfers between the electrical signal and the listeners.

Electromechanical conversion of audio from electrical to sound waves in air involves a number of variables which are governed by the choice and design of the speaker. But the final stage, acoustic transfer, occurs in the space between the speaker and the listener, and introduces major changes, especially in a large auditorium.

Fig. 5.6 represents a simplified auditorium scenario. In fact there are many reflections or echoes from the ceiling and floor as well, plus continuing reverberation which makes up the acoustic climate of the auditorium. All these reflections mix with the direct sound and arrive at

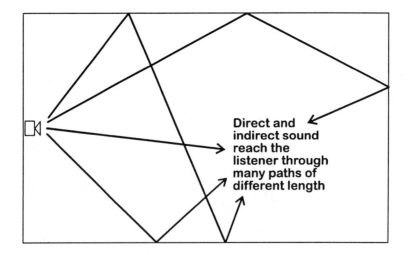

Direct and indirect sound reach the listener through many paths of different length

Fig. 5.6 How an auditorium modifies the sound between a speaker and the listeners.

the listeners via paths of different length, so the audience hears many repetitions of each voice transient arriving at different times.

PREPARING FOR UNAVOIDABLE TRANSFER ERRORS

Reverberation is an essential part of a musical program, but it can compromise intelligibility in speech or dialog because repetition of the essential transient sounds in normal rapid speech masks their message. By its very nature, a hall is going to change the sound, and people giving a talk in such conditions learn to speak more slowly. Public halls are given acoustic treatment which reduces the reverberation time at most frequencies in the speech spectrum, and speakers with suitable dispersion characteristics are used as explained in Chapter 6, but dramatic dialog would still find all large venues seriously wanting, were it not for the **signal processing** techniques used to prepare it.

Original dialog also may contain a high proportion of multiple reflections as it arrives at the microphone, causing similar transfer errors to those in the auditorium. Chapter 8 details some of the processing used in the dialog channels of movie soundtrack mixes, but that is only the tip of the iceberg. Many other media use high levels of sophisticated processing, such as signal modification in telephone and communications systems. Hearing aids come in stereo pairs, corrected for the shortcomings of each individual's hearing. They are used in stereo because the two channels can extract and maximize a conversation from a room full of people if suitably matrixed and processed. The effect can be reasonably simulated in a monaural channel with directional microphones and appropriate signal processing. Just as picture enhancement is part of the photographic and video medium, so audio signal processing is used to get the best out of sound channels that are limited in one way or another.

Audio signal processing is not a compromise, but an enhancement performed in the knowledge that the end result, despite electrical, electromechanical, and acoustic transfer errors, will be more real than a program **transferred** at every stage with a flat response and unlimited bandwidth, just as it occurs in nature.

NETWORKS

Electrical networks play a major part in Audio Engineering. They are the circuits that enable us to change loudness, separate the different frequency bands, mix and split audio signals. They are the basic elements of **signal processing.** Networks also play a part in control of DC, for example, separating mains hum from power supply voltages, decoupling the power rail that supplies various amplifying stages, and distributing different voltages.

The **voltage divider** is a resistive network for selecting a proportion of a voltage. Fig. 5.7a shows a network that would be equally at home distributing low current DC to an amplifying stage, or as a **loss pad** in an audio signal circuit. **Note that the greater resistance always faces the input,** so that the ratio of the divider can be infinitely small without short-circuiting the source. Note also that the output load will combine with this network to form an extended network, so if the load draws much current, the attenuation may be more than calculated.

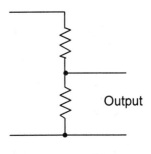

Fig. 5.7a **Voltage divider.**

The **gain control,** or **attenuator,** also sometimes called a **volume control,** is a variation on the circuit above, using a three terminal potentiometer (Fig. 5.7b). It appears in this form in the majority of cases because it is simple and effective, however, it does not present a constant impedance in either direction, so that the source load and the input termination on the following stage will vary as the sliding contact is moved up and down to change the output. In the AC power world, variable transformers are also constructed in this manner usually handling between one and five kVA (kilovolt-amps, or kilowatts if the power factor is unity).

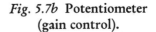

Fig. 5.7b **Potentiometer (gain control).**

A **rheostat** is a two terminal variable resistor. It is seldom used in audio because by itself it is not a network, and its action is therefore uncertain. Its

Fig. 5.8 **Rheostat.**

use is primarily as a current controller, whereas a potentiometer, as the name implies, controls voltage.

ATTENUATORS

Level reduction for component matching and operational gain control is achieved by using voltage divider networks with appropriate switching, or variable resistance elements. There are also electronic circuits and integrated circuit chips which have gain-control functions. Attenuators work at line level or below, where loss of small amounts of power or voltage is easily replaced with appropriate amplifying or gain stages. Fixed attenuators are called loss pads.

Attenuation at higher power levels, such as in speaker lines, is practiced in some cases, for example, to match levels or impedances in multi-speaker systems, or to include convenient volume controls in speaker circuits. But the use of resistive networks to match levels is very wasteful of power, as resistors get hot with the watts that were expensively acquired for driving speakers. Its not just a matter of loss; speakers also need close coupling to the amplifier because they do not give their best results unless they are well damped by its low impedance.

Transformers control speaker line levels better than resistive networks, with minimal power loss. Wire-wound potentiometers, even if designed to dissipate heat, are a poor reliability risk because of local heating at the contact point, and they should be avoided in favor of multi-tap switched transformers or line level attenuators.

Carbon track potentiometers have gained preference in most professional equipment over wire-wound or switched (stud) faders, due to advances in materials and quality of manufacture. The preference is also because transistor circuits are less particular about constant impedance terminations than vacuum tubes and transformers.

Carbon track potentiometers do not like DC, and quickly become noisy when leaky capacitors and other faults cause direct current flow through the sliding contact, moving or stationary. They are also prone to small drifts, especially if vibration is likely, and for this reason are often replaced by fixed resistors in critical alignments, following calibration.

Electronic attenuators include: remote volume controls, a whole range of effects processors and noise reduction systems, and limiters, volume compressors, and noise gates. Chapter 8 describes in detail the features and operation of the last three as adjuncts to recording and mixing, with emphasis on their operation and interaction in a signal processing chain.

VARIABLE ATTENUATORS

The potentiometer gain control is the device familiar to most users. It does not offer constant impedance to the **source**, or to its output load, but it has the advantage of being simple and cheap. The source sees the potentiometer as an impedance that is at its maximum near mid-travel, but the output load of the potentiometer sees a short circuit at minimum gain. For this reason it is imperative that the potentiometer should not be connected the wrong way round, because that would short circuit the source (Fig 5.7b).

Potentiometers are available in **linear, anti-logarithmic**, and **logarithmic** resistance increment. The last is used for most audio applications because the log characteristic makes a decibel-linear scale, and matches our hearing sensitivity to changes in loudness. Linear potentiometers will seem very crowded at the bottom of the scale, and near maximum volume they will appear to have little effect.

The term **gain control** suggests that something is supplying the gain, but it is obviously not the potentiometer. An active electronic module containing a potentiometer is a variable gain system, but attenuators in a line can only produce loss. When set for maximum output, gain controls

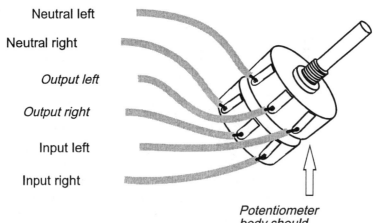

Neutral left

Neutral right

Output left

Output right

Input left

Input right

Potentiometer body should be grounded

Fig. 5.9 Connections to a stereo 20k + 20k logarithmic potentiometer.

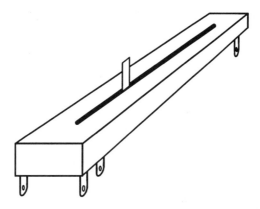

Fig. 5.10 Stereo slide potentiometer used as a mixer fader.

have an **insertion loss**, because they place resistance across the line, the exception being the rarely used rheostat, which would depend on a following termination to work at all. Insertion loss at full gain is due to the shunt resistor component of the potentiometer.

Potentiometers are made in both rotary (Fig. 5.9) and slider (Fig. 5.10) types. Step type, or 'stud' faders use switched **loss pads** of 1.5 or 2 dB per step, and have constant impedance in both directions. An advantage of stud faders is their precise attenuation repeatability and long life, although they can require maintenance. But because of increasing demand and improved materials the carbon track fader is the most common type used in mixers and other equipment. Electrically, it can be the same as the rotary potentiometer shown in Fig. 5.9, but there are also constant impedance types which use multi-section ganged potentiometers for each channel.

The need for constant impedance is associated with faders in a 600 ohm line in order to maintain correct termination for the incoming and outgoing lines, but within an integrated audio module, like a mixer, the system of distributed amplifiers and bridging inputs makes it practical to use simple voltage divider type attenuator circuits.

LOSS PADS

The variable attenuator has been detailed first, because it is one of the most visible components in an audio system, but the fixed attenuator, or **loss pad**, is a useful audio interface component, and is similar to the basic

85

networks which play a major part in any electronic circuit. For example, users of microphones will benefit by carrying 5 and 10 dB pads fitted inside connector pairs for insertion into a line. The application of loss pads, in the context of this book, is for audio level interface between electronic equipment modules (see Chapter 4, Noise Thresholds).

While it is possible to insert a potentiometer anywhere that level reduction is needed, in practice it will be found that hanging a cluster of pots (potentiometers) off parts of an installation is not only expensive and cumbersome, but it is a hazard waiting to go wrong. However, a pot can be useful for determining the resistance values.

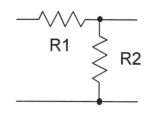

Fig. 5.11 **Unbalanced L-pads present line impedance at the input.**

There is one request that should be made of anyone fitting a pad inside a connector; that the connector or cable should be labelled to indicate its presence, because there is nothing more frustrating than chasing a loss that is not expected to be there. When making a padded cable, it is often convenient to put the series resistor at one end and the shunt at the other, to utilize available anchor points in the limited space.

The table on page 87 gives resistor values for a useful range of pads. Very large attenuations can cause high frequency loss in some circuits, but greater attenuation can be safely provided using several sections in cascade. Values within ten percent will suit most non critical applications. In line level audio, quarter watt resistors will generally handle anything the source can deliver. Pads are not used in speaker lines; transformers are used instead, as they do not waste power.

The pad and its wiring should be effectively shielded if the signal is less than line level, or if there is RF floating around in the form of radio or video signals, or hash from power mains switching transients or a square wave inverter. Even a short piece of wire has enough inductance to pick up radio frequency interference.

FILTERS AND EQUALIZERS

Networks are also used as frequency band boost and attenuate circuits that come in two forms. **Active filters and equalizers** consist of frequency selective networks in amplifying circuits. Active equalizers often operate via the negative feedback line. Since negative feedback reduces

dB loss	200 ohm Mic. Pads		10,000 ohm Line Pads	
	R1	R2	R1	R2
1	22	1639	1087	81967
2	41	772	2057	38610
3	58	484	2923	24213
4	74	342	3691	17094
5	88	257	4376	12853
6	98	201	4987	10050
7	101	161	5534	8071
8	120	132	6019	6614
9	129	110	6451	5501
10	137	93	6837	4625

Loss pad resistor values in ohms which can be linearly extrapolated to suit other impedances (Fig. 5.11).

Details of resistance substitution switch-boxes and the formula for resistors in parallel on page 92 will assist in choosing and making up unavailable resistor values.

---

gain, the network characteristic employed in a **negative feedback equalizer** is inverse to the function required; a bass boost negative feedback equalizer calls for a bass cut network in the feedback line, and so on. Multiple filter sections separated by buffer amplifying stages are used to make the slope of active filters as steep as required.

The other form is the **passive filter** or **equalizer**, which attenuates the appropriate frequency band both when it is cutting, and also when it is boosting. **Filters** act to positively remove part of the spectrum, but **Equalizers** permit passage of the whole spectrum, with level changes to some frequency bands.

Filters come in five types; **low pass**, or HF cut, **high pass**, or LF cut, **band pass**, which admits only a portion in the middle of the spectrum, **band reject**, the opposite to bandpass, and **dip filters**, which remove a narrow slice from the spectrum. The distortion meter characteristic of Fig. 5.5 is an example of a highly selective dip filter.

A sixth filter type, the **all pass filter**, is not directly concerned with attenuating unwanted parts of the spectrum, but passes all frequencies, attenuated only by its insertion loss. Its job is signal delay, correcting phase shift, or acting as a buffer around attenuating filter networks. The coaxial transmission line shown in Fig. 3.12 is an example of a multi-section all pass filter.

The filter networks shown in Figs. 5.12a and 5.12b are single section RC **(resistance-capacity)** networks, and have a slope of around 3 dB per octave. Steeper slopes can be achieved with multi-section filters. Insertion loss may eventually increase to the point where amplification is needed to recover it. This situation calls for an active filter, where feedback determines the slope of the filter.

The other way to increase the slope of a filter is by the use of inductors to replace the resistors in the voltage divider half opposite the capacitor. LC, or **inductance-capacity** networks make possible the design of passive

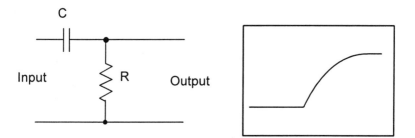

Fig. 5.12a **High pass single section RC filter and response curve.**

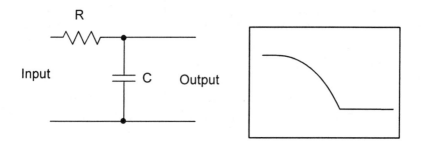

Fig 5.12b **Low pass single section RC filter and response curve.**

filters and equalizers with high slope characteristics. Although LC network calculations are outside the scope of this book, many publications on the subject exist, and readers with ability in calculus will find the study rewarding.

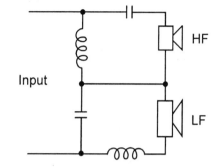

Speaker crossover networks are an example of LC networks, as demonstrated by the two-section design in Fig. 5.13.

Sometimes, the action of a filter is so subtle that its effect is inaudible, unless a spectrum analysis is made, for filters are often used to trim the nuisance ends of the spectrum, outside the normal range of the program. The

Fig. 5.13 **Single section speaker crossover network.**

use of both filters and equalizers usually produces a measure of phase shift, and unnatural sound can result if filters are too sudden, or if adjacent controls of a graphic equalizer are set to extreme opposites. On the other hand, an imperfect filter can often improve the way program sounds. Experience has shown that perfectly smooth filters, without ripple in the pass-band, can produce band limited sound that is dull and lifeless, but that a small peak generated by the filter, just before cut-off, can enhance the sound, making up for the loss of certain frequencies, and putting 'bite' back into sound of limited response. This type of response is called **passband ripple**, and filter designers may go to some trouble to eliminate it, unaware of its subjective value for certain uses.

The examples of basic equalizer networks shown in Figs. 5.14a and 5.14b work in the same way as the filters, but have an extra resistor, R_2. They are single section networks, still recognizable as basic voltage dividers.

The capacitor C determines the part of the spectrum where attenuation will occur, and the resistor, R_2, sets the attenuation limit. R_1 functions as the other half of the voltage divider which makes frequency band separation possible. A passive boost equalizer attenuates the frequencies that do not require change, leaving untouched the band it is boosting. The insertion loss of a passive bass or treble boost equalizer is equivalent to the maximum amount of boost the equalizer provides, plus its insertion loss. Bass and treble cut equalizers occasion only their network termination at frequencies that are unaffected.

Having obtained an understanding of the principles of loss pads, filters,

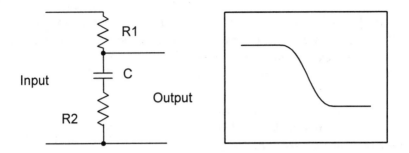

Fig. 5.14a LF boost, or HF attenuation equalizer and response curve.

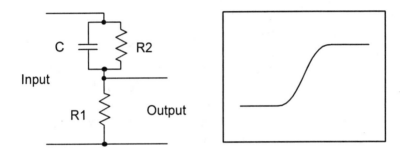

Fig. 5.14b HF boost, or LF attenuation equalizer and response curve.

and equalizers, how do we now apply the information. It may be that it merely helps to know approximately how commercial equalizers work, in order to facilitate their installation and use. All knowledge has value, and if it helps to give technicians confidence in the face of new equipment, that is good. But in this business how can we know when a challenge is coming up which will test our memory and ingenuity.

Although filter and equalizer design requires in-depth calculation, sufficient information is included here to establish the concept, and enable technicians to make useful networks by the trial method. Single section filter and equalizer networks are voltage dividers, like the one in Fig. 5.7a, but they contain frequency selective components. Component values can be found by calculation, but use of capacitor and resistor substitution switchboxes will avoid the need for complex calculations that are a less than essential part of field work. I hasten to assure readers that once tried, very little practice is required to make up simple correction equalizers, given access to an audio oscillator and an AC millivoltmeter or digital multimeter.

The magnitude of the capacitor determines the frequency band at which change occurs, and the resistor is chosen to determine the amount of change. The following table gives a regular sequence of values for single 12 position switches with which to make value substitution boxes. However, as with attenuators, the effect of filter and equalizer networks is easily altered by external circuit conditions and loads. Just because it works on the bench doesn't mean it will do the same thing when installed in its intended circuit. Self-contained commercial active equalizers and filters have buffer circuits or amplifiers at both ends for just that reason, to prevent the terminations offered by other equipment from interfering with the calibration.

Use substitution switch boxes to find suitable capacitors and resistors by sweeping the various values, noting the effect by frequency response measurement. Either a signal generator and AC millivoltmeter will measure a series of increments, or a **Pink Noise Generator** and **Spectrum Analyser** (Auditorium Calibration, Chapter 6), will display the frequency response as a graph.

Position	Resistor value in ohms	Capacitor value in microfarads
1	470k	.56
2	220k	.33
3	100k	.15
4	47k	.1
5	22k	.056
6	10k	.033
7	4k7	.015
8	2k2	.01
9	1k	.0056
10	470	.0033
11	220	.0015
12	100	.001

Values for component substitution switch-boxes.

RESISTANCES IN SERIES AND PARALLEL

Resistors are made in a range of **preferred values**, spaced in geometric progression, but fractional values can be made up by trial, with the nearest higher available value and another resistor of much higher value in parallel, say, ten or fifty times the value of the first. A resistor switch-box and an ohmmeter makes finding specific combination values easy, although they can be found by calculation. Resistors in series simply add; identical resistors in parallel have a total resistance equal to the value of each one divided by the number of resistors. However, non-identical resistors in parallel add as their reciprocals:

$$\frac{1}{R_{TOTAL}} = \frac{1}{R_1} + \frac{1}{R_2} + \frac{1}{R_3}$$

WHEN TO USE A FIXED EQUALIZER

For years, technicians were faced with a bewildering assortment of devices like phono pick-ups and tape heads that often had to be coaxed into calibration; they were not all standardized by any means, and could have losses which needed correction as well as the response curves designed for them. That seldom happens now, but the occasion will arise when a microphone sounds boomy and unintelligible, and the prepared technician can save the day, and a second visit, because there is a suitable capacitor and resistor in the tool kit.

Installation customers sometimes inadvisedly choose their own microphone, and many installations lack method; using a microphone without bass cut in combination with with full range speakers. So field technicians are occasionally called on to fix problems not of their own making. If this scenario comes up only once every five years, it is too often, if the means aren't there to remedy it.

Inevitably, this section on networks has to detail some eventualities which are more fully discussed in later chapters, but one of the facts of life is that equalization has to occur somewhere in a microphone-to-speaker chain in a normal auditorium system. As has been said elsewhere, but is worth repeating, voice reproduction from **full range** speakers can sound oppressively bassy because the perspective of close speaking into the microphone makes it so, and the larger-than-life speaker output will reverberate off every wall, favoring the bass end of the spectrum, because many auditoriums have less absorption in that region, and therefore longer reverberation times.

Measures to combat this include: choosing microphone or speakers with a tailored bass response, using the music/speech switch available on some microphones or amplifiers, adding a dialog equalizer to the system, or using the equalizers present on the mixer, if if one is used.

When nothing like this is available because of the simplicity of the installation, and both microphone and speakers have full frequency range; that's when to devise a simple equalizer network to insert in the line-out of the microphone preamplifier.

On the occasions where an integrated system has no independent microphone line output, the LF attenuator circuit of Fig. 5.14b will reduce bass response with a small insertion loss at other frequencies. However, if if is intended to boost the high frequencies, it is not practical to make the correction at microphone level because the higher insertion loss of the boost equalizer will move the limited microphone level down towards the noise floor. Therefore, rather than incur a substantial insertion loss in a microphone line, it should be done at line level. Bass cut would in any event be the first choice.

CAUTIONS REGARDING EQUALIZER USE

There is a basic rule in equalization. It is preferable to consider frequency band attenuation before boosting. Always boosting can result in an unnatural spectrum, as well as risking overload at some point in the amplifying chain. This comment particularly applies to the use of **graphic equalizers**; devices that can be more trouble than they are worth if not used with reserve. Undesirable phase errors degrade sound in a subtle way, causing, for example, some musical instruments to change balance in an orchestra, or lose clarity and character. Additional information on the use of graphic equalizers appears on page 130.

Despite this caution, equalizers should not be seen as instruments of phase error, because while they may produce phase shifts at the points of change, the effect of response correction by this means very often corrects the phase error that is associated with the microphone, speaker, or situation that is responsible for frequency discrimination. For example, in cases where process equalizers are used by design to produce a flat acoustic response from speakers, the result is usually better both in frequency response and phase response, the uncorrected device having mechanically generated phase errors before equalization.

RECOVERY AMPLIFIERS

Active equalizers and filters are made in two ways. Either a series of **passive** networks are followed by buffer and gain recovery stages, or else amplifying stages are frequency-controlled by negative feedback networks carrying the inverse equalization characteristic.

Passive equalizers you make yourself can simply be followed by recovery amplifier modules built on to an open printed circuit board. These are available generally from electronic and audio accessory distributors, and should be capable of recovering 10 to 20 dB. They use either an IC (integrated circuit) or a couple of small signal transistors, and have the same requirement as the low level microphone line driver described on page 63, where the purpose is to get the line level above the noise threshold. The IC version will probably have two channels for stereo or mono use. Instead of battery operation, the PCB (printed circuit board) can be powered from the nearest source of 12 to 18 volts that has a common negative ground. If space is at a premium, you may choose to make your own recovery or header amplifiers. I recently had to construct some two-transistor specials on the smallest PCB I could make, when the line from 70mm magnetic film heads in a new installation picked up hum from a nearby projection lamp transformer.

Since only a few were required for the job, the circuit tracks were hand drawn on the copper side of blank circuit board with a Dalo pen, which leaves a track of bitumen to resist the etching solution. The PCB assemblies were small enough to fit inside the magnetic head connector covers, and brought up the line level sufficiently to make the very high induced line noise insignificant. Sometimes, no amount of searching or expense will provide a specific item, and it just has to be made by hand.

Other losses calling for recovery amplifiers include mixing and bridging loss, splitting and distribution, and long line loss. Long transmission lines, both coaxial and twisted pair, have signal and frequency response degradation due to series resistance and inductance, and shunt capacity. Long telephone lines have repeater amplifiers at intervals to recover signal loss, and they also apply inverse equalization to restore high frequency losses, finally acting as low source impedance drivers for the following section of the line. As a succession of losses and gains balance each other in a mixer, line, or equalizer, the signal at all parts of the audio spectrum is maintained a safe distance from the noise floor so that it is fully recoverable.

ACTIVE CROSSOVERS AND SPEAKER PROCESSORS

The passive crossover network of the type shown in Fig. 5.15 has been used since the inception of the theater speaker, and is still used for domestic and professional speaker systems, but active crossovers have also been in use for several decades. The electronic crossover operates at line level and distributes two or more frequency bands to a separate power amplifier for each speaker.

Advantages of bi-amping are:

- There is no insertion power loss.
- Better phase response around the crossover points.
- Simple addition of active equalization.
- Speaker line sensing limiter in some systems.

As well as active equalization for bi-amped speakers, several speaker models of different brands, which have either no crossover network or a passive network, include an active equalizer as an essential part of the speaker, often termed a speaker process equalizer. Both active crossovers and process equalizers work at line level, and are placed immediately before the power amplifiers. If separate limiters or graphic equalizers for auditorium or room acoustics correction are used as well, they go before the processors because the latter modules are part of the speakers, and

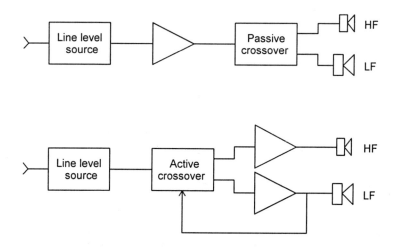

Fig. 5.15 Passive crossover (top) and bi-amp system. Notice the ease with which a speaker line sensing limiter can be incorporated in the active network.

should always be the last items in the signal path before the power amplifiers.

When separate power amplifiers are used with a mixer or preamplifier, the active crossover or process equalizer is inserted in the line between the source and the amplifiers as shown in Fig. 5.15. However, if an integrated amplifier or power mixer is used, then the processor is connected at an insert point. Insert points are brought out on a group of normalled jack sockets, or phono sockets with jumper connectors as shown in the following diagrams (Fig. 5.16). The normalled jacks will continue the circuit when the last jack is removed, but if the processor is disconnected from the phono socket group, then the left and right channel bypass jumpers will have to be re-inserted to return the system to normal. In the absence of jumpers, an ordinary short stereo cable will do the job.

Power mixers, which are mixers with built in power amplifiers, and integrated amplifiers like most stereo systems, don't always have insert points. It is possible to fit them if there is no alternative, but as discussed in Chapter 4 on page 69, the level at which the insert point operates is

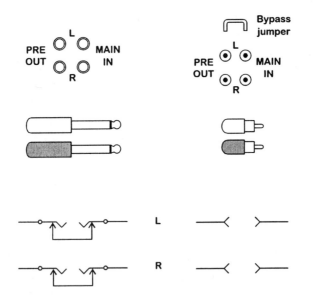

Fig. 5.16 Insert points on an integrated amplifier or power mixer, showing schematic diagrams of the open connectors that require bypass jumpers in the absence of the processor, and the self-normalling jacks. Alternative terms are INSERT SEND and INSERT RETURN.

critical for reasons of noise and distortion. In other words, the insert point signal voltage must be what the processor is designed for.

Commercial equipment should always be modified with caution. When a signal line is intercepted inside an integrated equipment module of any type, various conditions may complicate the situation:

- Buffer amplifiers or networks may be needed to correct level and match impedances.
- There may be a need to isolate or preserve DC conditions or sensitive circuitry.
- One or more feedback loops may exist across the circuit being interrupted.

Circumstances do arise where it is just not possible to use an outboard processor with certain equipment. In the case of stereo integrated amplifiers, the tape recorder output can be used to drive the processor while the signal return is made via the recorder playback input. Under this condition, all signal to the speakers comes via B-monitor, Monitor 2, or the tape replay monitor.

It is not possible to do this with all integrated amplifiers. For example, a five channel AV or Home Theater amplifier will only operate in stereo mode from Tape Monitor, and when all five speakers are operating, left and right channels will not be equalized.

Purchasers of power mixers or multi-channel integrated amplifiers who intend to use graphic equalizers, speaker processors, compressors, limiters, or active crossovers should be sure the equipment will do what they need, as modifications to commercial equipment, while generally possible, are expensive, somewhat risky, and will definitely spoil the resale value of the system.

Chapter 6.

Speakers

SPEAKER POWER RATINGS

One of the prime factors in choosing a speaker is the power range in which it is required to operate. Amplifier power applied to a speaker is greatly misunderstood, so power rating needs definition. The real limit of a speaker's power handling capacity is the level at which it starts to produce a non-linear output. Even if it will survive a higher drive level without damage, there is no point in pushing it beyond the level where it starts to sound bad.

Most audio program, the real sound we listen to, has an average power representing about 70 percent of its peak value, so it is clear that the method of rating the speaker has to be stated. Without more exactly specifying the type of signal, a 60 watt rating, for example, suggests that the speaker would probably be safely driven with a continuous 35 or 40 watts. Any treatment of program material that gives it an abnormal spectrum reduces a speaker's power handling capability by an equivalent factor. 6 dB of bass boost or volume compression will reduce the speaker's effective rating by 6 dB, because the spectrum is then abnormal by that amount. In professional audio, the need for substantial bass output is dealt with by adding bass speakers; not by turning up the bass on the full range speakers.

Volume compression, limiting, and clipping moves the average level up towards the peak value. A program so treated is outside normal program specification, and may be damaging to speakers in that form, since continuous application of signal can represent a condition similar to square wave or direct current. Square wave is damaging because the voice coil is unable to follow the peak level high frequency component, and also because the drive is continuous, without intervals. Like all electric motors, speakers can't follow the rapid acceleration and direction changes occasioned by a high level square wave signal, plus continuous operation beyond the design duty cycle: the drive to rest ratio demonstrated in Fig. 6.1.

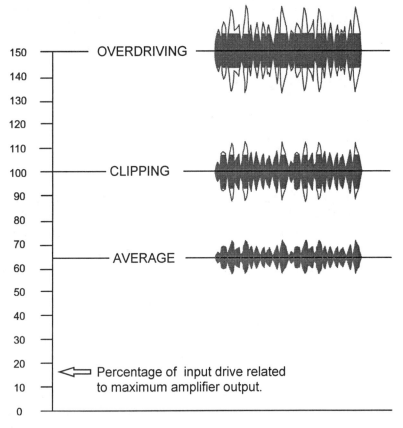

Fig. 6.1 How *Square Wave Distortion* damages speakers.

The waveform diagrams show that the average power of program signal is about 35 percent below its peak value, and that when the amplifier's clipping point is overtaken by the signal peaks, waveform compression increases the program's duty cycle out of proportion.

Because the waveform shape change caused by amplifier overload clipping forces the RMS value of the output upwards towards the peak value, amplifiers driven into distortion can produce a total energy output higher than their power rating.

Most systems have more than enough gain to drive power amplifiers well into distortion; twelve o'clock on the volume control would be about the limit for most integrated amplifiers, or mixers plus power amplifiers with normal program input levels. The danger of using a small domestic amplifier, say 20 watts, to do a big job like a get-together with twelve people, is that few of them will notice a bit of distortion under the circumstances, but the 100 watt speakers will be quite dead the following morning, the voice coils having cooled down after the bonding materials melted and glued the whole assembly out of alignment, even if the speakers are not actually open circuit.

IT'S NOT THE POSITION OF THE VOLUME CONTROL THAT COUNTS, BUT THE VOLTAGE AND WAVEFORM FED TO THE SPEAKER.

ESTIMATING THE POWER TO A SPEAKER

In professional audio, where the effect is similar but more expensive, it is not always obvious just how much power is being fed to the speakers, especially bass speakers.

Connecting an AC voltmeter to an 8 ohm speaker line will tell the operator that when 40 volts is displayed, the average drive, by calculation, is 200 watts, plus at least another 6 dB for peaks that the meter movement can't follow. But that's only approximate, and operators don't want to work with their eyes glued to level meters, so a more user friendly method is to use a speaker line sensing limiter before the power amplifier, or better still, use an amplifier that has an input-output comparator coupled limiter, which won't allow the amplifier to clip, but will hold the output within its rated maximum. All the operator has to do then to safeguard the speaker, is choose an amplifier of suitable maximum output, and make sure the program has normal attributes.

IEC SPEAKER POWER RATING

In order to standardize speaker power ratings, the IEC (International Electrotechnical Commission) power rating method specifies a test which drives the speaker with **pink noise** for 100 hours. The highest level in watts at which the speaker can safely operate for that period becomes the speaker's IEC Power Rating.

The pink noise used in this test, and also for speaker and auditorium calibration, is a broad spectrum test signal which has equal power per unit bandwidth, and is described more fully later in this chapter under Auditorium Calibration.

Speakers are expensive and inherently valuable for their performance capabilities. They should be chosen with adequate capacity for the job they are to perform, in much the same way as a car engine. Lend your speakers only to people with a reasonable level of of power rating knowledge.

DIRECTIONAL PROPERTIES AND STEREO IMAGE

The way sound travels depends on its place in the audio spectrum. High frequencies radiate from the source at an angle determined by the geometry and size of the source, reflecting off surfaces at the same incident angle as they approached; like light from a mirror.

To make a realistic reproduction of an original sound, the speakers' angles of propagation, frequency range and phase relationship should resemble those of the original sound source. So they might aim at recreating a fair simulation of all the multiple direct and reflection paths of the original, at all the frequencies in the audible range (Fig. 6.2).

Experiments with different speaker types indicate that wide angle mid and high frequency propagation is an essential factor in production of an

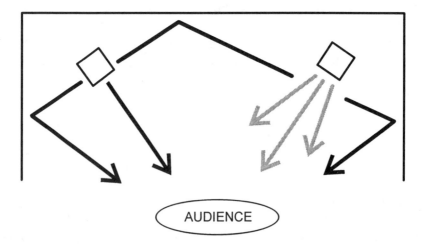

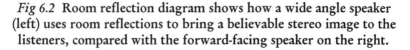

Fig 6.2 Room reflection diagram shows how a wide angle speaker (left) uses room reflections to bring a believable stereo image to the listeners, compared with the forward-facing speaker on the right.

effective stereo image. Manufacturers have produced speakers with dispersion patterns from 180 to 360 degrees because wide angle dispersion is able to make a believable stereo image.

What does it matter if all the reflecting paths and phase complexities are not exactly as they were in the original sound, and could we tell the difference anyway? The speakers, and the room they are in, **invent** much of the image, but the **impression** they create has more value than truth. It's like going to the movies and undergoing a Cinematic Experience so real that we become absorbed in the scene and the characters, in the full knowledge that some voices may be dubbed, the sound effects added later, and that parts of the picture may be computer generated images, backdrop murals, or paintings on glass.

Original sound sources of any size, like orchestras or bands, radiate different sound in all directions, reflecting off walls and ceiling, and to faithfully reproduce such complexity is impossible and unnecessary. Speaker systems are, however, called on to make a fair simulation in a home situation, and they do it by propagating the sound through a wide angle and utilizing reflections from walls and ceiling to simulate the multiple sound paths of the original. Very often the best stereo image is achieved with the speakers at the narrow end of a room or studio instead of across the longest wall.

STEREO SURROUND

The other way to make a realistic multi-directional sound field is to matrix the two program channels into several more, as is done with home theatre and cinema surround systems. The system selects components of left and right channels which are identical in phase, and directs them to a Center channel. The remainder is directed to Left and Right. In addition, a pair of surround channels is created from the **difference** between the original program channels, and separation enhanced to the extent that no center channel sound appears in the surround information.

Although designed for reproduction of Stereo Motion Picture Soundtracks, Surround Decoders work well with many real stereophonic programs, but not with monaural material presented in simulated stereo. No actual **encoding** exists in stereophonic recordings, yet through a surround "decoder" and speaker system, all components of the original program can assume their proper place. Left, center, and right spread across the virtual stage, and individual instruments can be located in depth as well as laterally. The ambience of the concert hall surrounds the listener as off-center chorus voices, percussion, and second instruments appear to the side, suspended between the front and rear of the sound field.

Extracting the center and surround information is not the only way to hear the surround channel. Some wide angle speakers, suitably positioned, are especially good at it using only two channels, and can generate rear channel information in the room by reflection.

On many occasions pure stereo concert hall recordings don't sound as good played in surround mode; voices seem far away and soloists drift in and out of center channel. The suggestion is that whichever mode suits individual programs is the one to use; stereo or stereo surround.

If you are assembling a surround system for total theatre, that is, movie soundtracks and all stereo music and TV entertainment, use full range speakers for your left and right channels; wide angle speakers won't be out of place. A graphic equalizer on center channel helps to make dialog sound more real.

POOR MAN'S SURROUND

To demonstrate the simplicity of the basic surround sound principle, a setup like the one shown in Fig 6.3 will produce a surround channel quite well, but full decoders have three attributes in the surround information that this method does not: A high degree of front channel rejection, noise

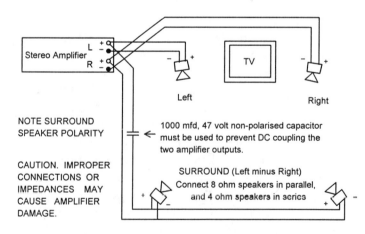

Fig. 6.3 Simple surround channel decoder.

reduction expansion, and bandpassing at about 90 and 7,000 Hz. But it can produce a convincing surround channel, costs almost nothing, and gives an extra ambience dimension to stereo music. A description of Home Theater systems appears in Chapter 15.

USING SPEAKER DIRECTIONAL PROPERTIES

Wide angle dispersion has its place, but there are many situations where limited angle sound propagation is needed; to control reflections in reverberant venues, to direct the available power where it is needed, and to enable multiple speaker arrays to operate without mutual interference.

Domestic speakers have intentionally wide dispersion patterns, and are mostly intended to operate in a room where the walls reflect the sound and set up a broad field. Used outdoors, their performance is poor because the sound, especially the bass, dissipates and is not reinforced by reflections.

Professional speakers are more efficient and handle more power, either individually or in distributed arrays of small speakers, and the propagation pattern they are chosen for is specific to the requirements of the job (Fig. 6.4).

In the larger power speakers, output is directed into a fixed audience area, and spill outside this area reduces sharply in the mid and high frequency ranges, the intention being to minimize unwanted reverberation by keeping stray sound off the walls, and to keep spill away from

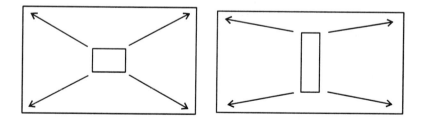

Fig. 6.4 **A horn speaker or direct radiator array (left), propagates sound in the direction of the horn mouth or angled driver cones. Horizontal dispersion angles are from 90 to 120 degrees. Column speakers (right) or** *line radiators,* **have a wider horizontal pattern** *perpendicular to* **the orientation of the column, but have less sharply defined dispesion angles than horns.**

local residents in outdoor events. In covering less specific venues, where the listeners move around, or where the area is very big, numbers of small distributed speakers may be used, with somewhat reduced intelligibility where the program contains dialog or speech.

The choice of direct sound from central clusters of large speakers is preferable to the use of distributed small speakers, because the latter method loses clarity due to multiple length sound paths to the listeners. In venues where centrally located speakers are impractical, delay units are sometimes used to time-align the separated speakers so that sound from all the sources reaches the majority of listeners simultaneously.

Professional speakers are designed to maintain the same frequency response over their propagation angle, enabling the speakers to be used successfully in matched arrays to cover a specific audience area. To achieve wide angle coverage, speakers are mounted in a cluster to make a continuous horizontal spread with minimal overlap as shown in Fig. 6.5, while suitable types of direct radiator speakers can be be stacked in a vertical line to form a column, which has wide horizontal dispersion. Speakers already designed as columns, or the horn plus low frequency driver type illustrated below are not suitable for stacking into columns.

When using multiple speakers placed beside each other, overlap of adjacent patterns causes interference which partly cancels some frequencies at positions between the speakers, giving a "phasing" effect as the

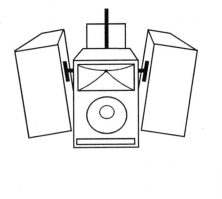

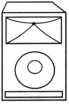

Fig. 6.5 A suspended cluster of four 120 degree horn speakers: alternatively, a pair spaced apart, facing forward.

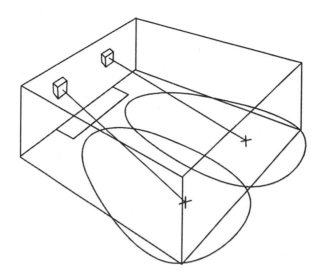

Fig. 6.6 Declination and relative rotation angles of two speakers
covering an auditorium.

listener moves across them. Interference between adjacent speakers requires them to be either laterally separated by not less than six feet,
rotated away from each other, or both, until overlap is minimized and
continuous coverage established. Examples are shown in the suspended
cluster of Fig. 6.5 and the auditorium perspective in Fig. 6.6.

Depending on the height at which they are suspended, speakers are
angled down to cover the required area, the aim being to provide the same
Sound Pressure Level (SPL) for every listener. In order to do this,
speakers mounted high are pointed down, over the heads of the front
rows of seats to a position about three quarters towards the back of the
auditorium, while speakers on portable stands are directed horizontally,
but they should be not less than six feet above the listeners' heads.

One of the prime rules in speaker placement concerns the use of
separate bass speakers. When the spectrum is divided between high and
low frequency speakers via either passive crossovers, or the program is
Bi-amped using active crossovers before the power amplifiers, it should be
attempted to propagate the same frequency response all over the audience
area. This will be achieved by making sure that the bass speakers are the
same general distance from the listeners as the high frequency speakers. A
percentage of listeners will experience a disturbingly unbalanced spectrum
if the bass speakers are too close compared to the mid and HF units.

Bass speakers are often suspended at the same height as the other units, but with two differences:

1. Bass below 200 Hz is essentially omnidirectional, and the bass speakers do not have to be pointed. But if it makes an installation look better, there is no reason why they should not be angled down.
2. Bass speakers should be located where there is an effective reflecting surface behind them, like a solid wall, as they do not have to be placed right near the high frequency speakers. Their position will not be detectable at the frequencies they handle. Figs. 6.8 and 6.9a show the theory and methods of improving bass propagation.

FREQUENCY RESPONSE OF SPEAKERS

Single cone speakers have long been used to cover the full frequency range. They work because the material of the cones is compressible, and when it is driven from the center and loaded by the edge suspension, it develops concentric areas that move independently at different frequencies.

Sometimes concentric cone corrugations are used, and doping with rings of various damping and stiffening compounds determines the way the cone "breaks up" to produce multiple adjacent frequency bands. Some great results are achieved by this method, often in multiple direct radiator arrays. It must be pointed out that there is no right or wrong way to make a speaker. All methods have merit, which depends very much on its purpose and the vision of the designer. Other designs use rigid cones and diaphragms with multiple drivers, effective in different frequency bands, and separated by crossover networks. One of the main reasons for this is to limit the band in which the various drivers have to operate, to optimize efficiency by operating each driver in its ideal frequency band.

In domestic speakers, gradual crossovers are often used to overcome phase problems; smooth transitions from one driver to the next can sound better in moderate power speakers, whose aim is to reproduce music as well as possible with minimum phase error. Steep crossovers, between 6 and 18 dB per octave are valuable in higher power speakers, as they give the drivers narrower bands in which to operate, saving power and voice coil heating, and permit the use of more specialized drivers. Steep crossovers are mandatory for high frequency horn loaded compression drivers, which have limited excursion, and can not handle low frequencies, being in any event unable to reproduce them with standard horns.

While the bass driver crosses over to the high frequency, or to the mid-range driver in a three way system, the crossover point may be anything from 100 Hz up to 2,500 Hz. The bass section measured by itself may show a flat response up to 6 kHz, but the reason it is not used at higher frequencies is because the job is done better by the mid and high frequency drivers which have better dispersion angles and produce more coherent sound with better transient response. Steady state measurements do not give the whole answer; a speaker's dynamic performance involves far more factors. For reasons of size and the **inverse square law**, every driver enclosure, or horn, has a best performance spectrum for efficiency, dispersion, and freedom from distortion and response errors of phase, frequency, and transients. In a professional speaker, if the bass driver is used up to its response limit before crossing over, clarity of sound right to the back of the auditorium will suffer, because its radiation pattern is non-ideal at those frequencies, and intermodulation distortion will also result through Doppler Effect (page 115). The choice of crossover from a low frequency driver is influenced by the speaker's intended use; 500 Hz or lower for dialog programs, and up to 1200 Hz for music and vocal programs.

RESPONSE LIMITS OF SPEAKERS

It is possible to make a speaker with almost any high or low frequency cut-off point, but the question as to what is desirable depends on its purpose. The upper limit of a high frequency driver is subject to several philosophies. Speakers used in **large auditoriums** are often designed with a roll-off above 15 kHz in order to bandpass the system and avoid upper harmonic imbalance caused by the higher than natural loudness of reproduction, and high frequency delays and cancellations due to multiple long-path reflections. On page 120, a simple design alignment is described which automatically sets the cut-off frequency of compression drivers. At the other end of the scale, speakers in **low budget domestic systems** may have limited bass, and in order to produce balanced sound, the high frequency range is also limited intentionally (630 Hz, center of the audio spectrum, page 54).

Conversely, in an **ideal listening environment**, high frequency reproduction of music and singing voices should extend past audibility, rolling off smoothly towards 25 kHz in order to take advantage of low phase distortion. We can't hear the highest frequencies, but if they are reproduced in balance, neither will we hear phase distortion due to rapid cut-off, and subjectively, it will sound better. Comparative listening tests indicate the validity of this theory in the naturalness that extended high frequency response exhibits, especially in small venues or at home.

Similar comments apply to the bass cut-off; a flat response that cuts off sharply can sound less natural than a bass response that tapers away smoothly below cut-off (3 dB loss point). However, the trend towards compact speaker design generally favors rapid bass cut-off.

The sub-bass region shown in Fig. 6.7 operates below the range of most music, covering the octave from 20 to 40 Hz. Usually housed in a separate enclosure and driven by its own monaural low pass filter and amplifier, a Sub-bass speaker is a large and expensive extra in systems that don't need it, but is useful for extending the bass range to include the lowest fundamental frequencies that pipe organs and a few acoustic orchestral instruments produce.

Sub-bass is a frequently misused term, and is often incorrectly applied to the separate bass module of a speaker system of normal spectrum. "Sub-" means **below normal**, and represents an extension of the generally accepted lower limit of about 35 Hz, whether or not different parts of the spectrum are handled by separate speaker boxes.

A sub-bass channel is very appropriate in cinemas, where great use of this part of the spectrum has been made since the advent of stereo sound-tracks, but it is inappropriate in a disco environment where the

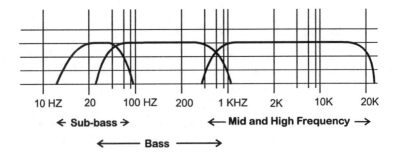

Fig. 6.7 Exponential chart showing typical crossover frequencies, and indicating the sub-bass region, the extra octave needed for some types of program.

spectrum of useful sound ends at 40 Hz, and will only waste amplifier power and clog up the environment with unnecessary subsonics, which are a serious discomfort factor at continuous high levels. The sub seldom speaks except for an outstanding part of the program, and will be silent for most of the time. Sub-bass speakers and filters are described in Chapter 15.

STANDING WAVES

In any room or hall less than 30 feet long, original and reflected sound waves meet each other, adding and subtracting at wavelength nodes, and create zones of high and low sound level perpendicular to the direction of propagation. For example, 1000 Hz will form high level nodes 1.1 feet apart, with low level troughs between those; the speed of sound at 1100 feet per second forming sum and difference level nodes as each wave meets a reflection.

The effect goes largely unnoticed in normal program, except at low frequencies, where the wavelength is large, and can cancel much of a speaker's output at various positions in a small room. The lower the frequency, the more noticeable the loss of bass at various positions in a small room. The answer is often to strategically place bass resonator absorber panels, detailed on pages 182 and 183, to attenuate the reflection path; this makes the acoustic environment more like open air, and reduces the effect of standing waves.

Speaker placement is also a means to combat listening room faults, especially where the bass speaker is separate, and has a bearing on standing wave production (Chapter 15). Using a separate bass cabinet may help because it can be placed anywhere in the room without localizing its output. A method of ideally positioning bass speakers will be described in Chapter 15.

EFFICIENT BASS SPEAKER PLACEMENT

Professional bass speakers used outdoors and in large halls benefit from a solid background against which to drive the air, in the absence of enclosing walls (Fig. 6.8). Bass below 200 Hz is omnidirectional, and the percentage that radiates away from the listeners is not only wasted, it's a nuisance to the neighbors. In the absence of a backing wall, a heavy baffle wall of three quarter inch material, braced to the speaker box to extend

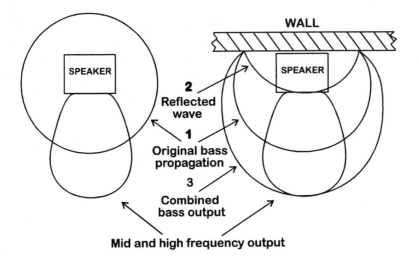

Mid and high frequency output

Fig. 6.8 Comparison of the *forward* bass output of a speaker with and without the reflection provided by a wall.

the width of the speaker (Fig. 6.9), will reinforce the bass and make it more directional, effectively extending the response. Because speakers are normally designed to work without this type of augmentation, the advantage in bass propagation may not have a flat response, so spectrum analysis and equalization is recommended following such additions.

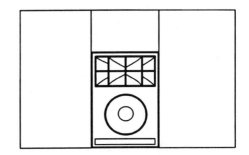

Fig. 6.9 Three-quarter-inch chip-board baffle around a speaker.

SPEAKER BAFFLES

In order to propagate bass frequencies from a speaker driver, the two sides of the cone must be separated by a baffle. Sound travels at 1100 feet, or 335 meters per second. When the cone (Fig. 6.10) moves in one direction, a compression wave travels at that speed around to the other side of the cone and cancels the corresponding decompression, so wavelengths longer than the bypass distance are not propagated beyond the confines of the cone. The baffle increases the distance the compression wave has to travel in order to cancel, and lowers the cut-off frequency that can be radiated away from the cone.

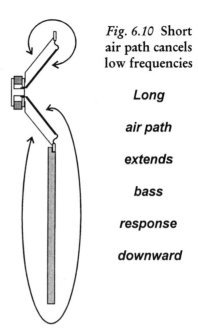

Fig. 6.10 Short air path cancels low frequencies

Long air path extends bass response downward

BASS BOX DESIGN

Better bass response is obtained by isolating the back of the cone in a closed box filled with damping material. The air mass in the box is chosen to correctly load the cone for optimum bass response; it acts as a weighted spring which balances the area, mass, and compliance of the cone. This enables it to handle more power and radiate an extended bass spectrum from a small cabinet. In the **air suspension** system, sound is radiated only from the front of the cone. But if the box is ported, efficiency increases due to use of the back-wave from the cone.

Figure 6.11 shows how the compressible mass of air in the box, driven by the back wave of the cone, produces **acoustic phase inversion** to radiate bass from the port, in phase with the front wave of the cone.

There have been years of research expended on speaker boxes, notably the ported box design. Though not representative of all domestic speakers, the "Bass Reflex" principle is almost universally used in professional speakers other than horns due to its efficiency and low distortion.

The bass reflex box can be tuned to optimum by refining the design parameters of cone diameter and resonant frequency, volume of air in the box, the ratio of electrical resistance to cone mass and suspension compliance, the area and length of the port, and the amplifier damping factor. Under these conditions, a ported enclosure can exhibit better power performance than an infinite baffle.

Sometimes the inside of the box contains damping material to reduce internal reflections, and therefore standing waves. Other designs increase cone damping by backing the driver with a tightly stretched curtain of damping material which acts as an acoustic resistance. Yet other designs place an acoustic resistance at the port, or use a "drone cone" instead of an

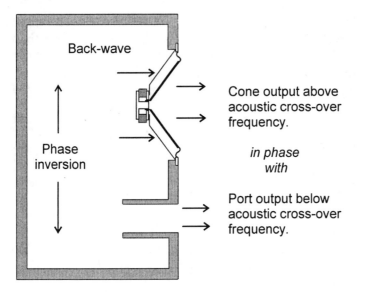

Back-wave

Phase inversion

Cone output above acoustic cross-over frequency.

in phase with

Port output below acoustic cross-over frequency.

Fig. 6.11 **Back-wave adds in phase to the front-wave of ported boxes.**

open port; a diaphragm driven by the air movement. In a conventional design, the length of the port shelf or tube performs a similar function.

A correctly designed ported box is an acoustic high pass filter which can be designed to maintain a flat bass response down to cut-off, below which point the output falls rapidly. A variation of this principle uses a smaller box tuned to an earlier but less steep fall in output. Bass boost equalization is then applied to produce a flat response with an appropriate low end roll-off rate (Fig. 6.12). Active equalization is used with some domestic and professional speakers. A few of these designs include a speaker-line sensing limiter for overload protection.

All other things being equal, cone diameter is not the major factor in extended bass response; the diameter of a bass driver is related to the efficiency of the speaker, and how much power it is required to handle, but the bigger the cone, the bigger the box needed to air-load it.

Very compact systems are designed down to the last millimeter; if any parameter is changed, performance suffers. Optimum box tuning lowers cone excursion to almost zero at resonance, to the point where the **Doppler Effect** is negligible because the lower bass is radiated by the port, not the cone, and this enables the same cone to radiate high frequencies as well. However, if a high compliance bass driver is used in an incorrectly designed box, or if the box leaks, the driver is not sufficiently loaded and can exceed its safe excursion and suffer damage.

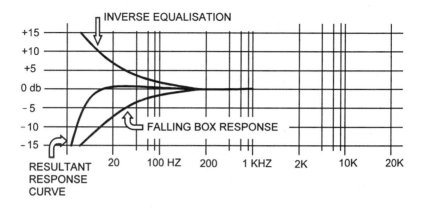

Fig. 6.12 A designed taper in the bass response of the box is corrected by the inverse boost curve, in this diagram representing active equalization to permit the use of a smaller bass speaker.

There is still justification for loose specification speaker box design where generous proportions can produce good results, but operation at high levels requires that the Doppler Effect has to be overcome by crossing over to a separate high frequency driver because the bass driver will have greater excursion. Efficiency of a speaker system relative to size is less important today in view of the ready availability of high power amplifiers.

INTERMODULATION DISTORTION

As discussed, the **Doppler Effect** plays a major part in choosing to cover the full audio range with one speaker cone, or cross over to separate mid and/or high frequency units. Doppler is the effect that causes sound to vary in pitch when it is radiated from a moving surface. Due to the relatively low speed sound waves travel in air, 1100 feet per second, a stationary observer hears lower pitch when the source is moving away, and higher when it is approaching.

In full range drivers which have large cone excursion at low frequencies, **intermodulation distortion** occurs. This happens because the cone is moving forwards and backwards at low frequencies, so high frequencies radiated by the moving surface will vary in pitch at the low frequency rate. To avoid intermodulation distortion where it is likely to be generated, different drivers handle high and low frequencies via a cross-over network, so that high frequencies are radiated from a relatively stationary surface.

PHASE RESPONSE OF SPEAKERS

To sound natural, a speaker should reproduce all frequencies in their original time relationship. Speaker phase error has various causes. It can be the result of rapid changes in response at different parts of the spectrum, whether caused by speaker cone breakup (independent movement of parts of the cone), steep filter networks, severe equalizer application, or multiple acoustic reflections. Users of Graphic Equalizers are advised to make spectrum corrections minimal and smooth. For example, swinging two adjacent frequency band controls to opposite extremes will produce an unacceptable degree of phase distortion, suggesting that it is better to under-correct than to achieve perfect frequency response at all costs. Phase error makes sharp transient sounds lose clarity, and subtle orchestral instruments disappear or lose their character. Test sounds used to verify good acoustic phase performance include short bursts of square wave which are picked up by a microphone and displayed on an oscilloscope. For judgment by ear, any signal containing a series of clicks, like tap-dancing, or a mix of low and high frequency square wave tones, is reproduced and heard from different positions. Some phase errors make the clicks run together indistinctly, and speech more difficult to understand.

Phase relationships in multiple driver speakers are subject to the mounting positions and types of HF and LF drivers, because the relative distance of the points of origin will have a bearing on when sound from each driver reaches the listener. Speakers have been designed with stepped cabinets to correct the driver positions. High frequency drivers are sometimes intentionally connected 180 degrees out of phase if optimum results are achieved by doing so. Both domestic and professional speakers have occasionally been designed in this manner. There is evidence that the equalization required to optimize a speaker's response also works in the right direction to correct its phase errors.

SPEAKER AND DRIVER PHASING

Phase has quite another meaning in a different context. We've been discussing phase errors of various amounts at different frequencies, which can be referred to in degrees of angle; the time-relationship between various parts of the complex wave that makes up an audio signal. But the following refers only to **signal inversion**, or electrical phase differences of exactly 180 degrees that affect the whole signal at all frequencies.

In order to properly reproduce a wave-front of positive pressure picked up by a microphone, a driver cone must move forward, towards the listeners. The first wave-front of a speech syllable will lose impact if it moves backwards, or it will be less than real. It is also important that all the speakers in a system should be in phase with each other, since bass propagation will be weakened by cancellation if one speaker sucks while another is blowing, to put it simply, and the smooth sound field of higher frequencies will be disrupted.

When assembling a speaker, new or under repair, it is necessary to ensure that all the drivers are connected in phase, because cancellations between signals generated in the same speaker box due to an out of phase driver will seriously compromise efficiency and frequency response. In the case of speakers containing arrays of identical full-range drivers, one out of phase will have insufficient air-loading of the cone, and it will tear itself to pieces through over-excursion, driven by the others. Speakers with cross-over networks between the drivers will have sections missing from the spectrum if any drivers are out of phase.

A simple test should follow all speaker repairs or assemblies, as it is more than easy to make an error. Just touch the speaker terminals with a nine volt battery. When positive voltage is applied to the positive terminal, the cones should all move outward, away from the box. Note that some speakers have drivers on the side or the back. They should all move outward together, away from the box.

Mid-range, high frequency drivers and compression drivers will not show perceptible movement with the phase test, so connections should be carefully checked against wiring colors. High frequency drivers are sometimes intentionally connected 180 degrees out of phase, and some brands of professional driver have forward cone movement with negative voltage on the positive driver terminal.

If there is doubt about phasing of the various drivers in a speaker, a pink noise generator, microphone, and spectrum analyzer will readily display the difference (Fig. 6.19 and pages 127–134).

SPEAKER EFFICIENCY

Horn loaded compression drivers offer the most efficient way to convert signal to sound, as the exponential horn multiplies the sound approximately by a factor of five, efficiently matching the short travel of the diaphragm to low density air at the horn mouth. Next in order of efficiency come direct radiator cone speakers. Last, but still worthy of consideration, are the variety of electrostatic and ribbon speakers which have long been used as a medium nearer the density of air than heavier rigid cone and diaphragm assemblies. The principles are not applicable to high power speakers, but prove themselves otherwise in natural middle frequencies, and smooth and extended high frequency response, in the endless search for natural sound reproduction.

The efficiency of conventional bass or full range speakers is, broadly stated, a matter of relative size. Comparable domestic speakers have been dramatically reduced in size, and therefore in efficiency, over the last three decades, countered by a four to one reduction in voice coil impedance from 16 down to 4 ohms. Since the DC resistance of voice coils is very close to rated impedance, lower resistance also means less heating loss.

Professional speakers cover a wide range of types to fit the various uses; installed or transportable, large venue or commercial, horn or direct radiator. The trend also moves continually towards compactness and improved sound quality, and with both parameters, the efficiency of speakers may be reduced. This is not necessarily a problem, because providing sufficient amplifier power is now relatively easy; the combination making for lower distortion and better frequency response. Comparable power amplifiers and speaker styles for top of the range home use rate as follows over the periods shown:

1950s and 60s: 5 watt single ended vacuum tube, and 10 watt push-pull vacuum tube amplifiers drive 8 to 12 inch 16 ohm speakers in 100 to 150 liter open backed cabinets or ported boxes, sometimes with a tweeter added to reduce intermodulation at high bass levels (refer to pages 115 and 116).

1970s and 80s: 40 to 100 watt stereo transistor amplifiers; 4 to 10 inch 8 and 4 ohm bass drivers in ported boxes with tweeters, and sometimes a mid-range driver.

1990s: As preferred amplifier power increased above 100 watts, the domestic market's preference grew in favor of compact speakers with high specification enclosures and small diameter drivers, designed to handle greater power.

THEATER SPEAKER EVOLUTION

In 1928 the first large auditorium speaker was demonstrated reproducing a stereophonic music broadcast in a public auditorium using a pair of two section exponential horn loaded speakers. When further developed it became the Motion Picture Theater standard speaker, using amplifiers of between ten and twenty five watts output.

In its established form it consisted of four eighteen inch low frequency cone type drivers serving two four foot high "W" bins, exponential bass horns crossing over at 350 Hz to a single multicellular high frequency horn (Fig. 6.13) driven by a compression driver with a three and a half inch duralium diaphragm (Fig. 6.15).

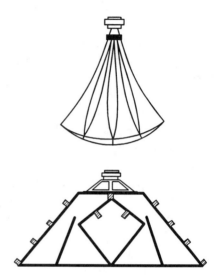

Fig. 6.13 **Plan view of HF horn and W-bin.**

As shown in Fig. 6.14, the whole structure was about ten feet high and the electro-magnets of early drivers were excited by a 25 volt, one amp power supply. The quality of Motion Pictures rivaled other sound media for two decades.

The lower frequency response of horns is limited by the length of the horn and the area of the horn mouth, in its relationship to the wavelength of the lowest frequency it can propagate. Baffle wings were added to the sides of the entire structure to extend low frequency efficiency. A similar principle is used today in stereo cinema installations in the form of a baffle wall in which the front channel speakers are mounted. The baffle wall gives power and impact to the low frequency component of sound effects, although the sub-bass speakers are not necessarily included in the baffle wall speaker line-up.

In theatrical sound, it is necessary for dialogue to radiate from a single source, because multiple speakers reproducing an identical program will reach parts of the auditorium through several different length paths, resulting in loss of intelligibility by repeating transients. To keep the dialog confined to a single source, higher powers were accommodated by adding a second driver to the HF horn via a Y-throat coupling.

The original high frequency horns were multicellular, and exhibited nodes of high and low level in the dispersion pattern. The voice coils were wound with aluminum ribbon on edge, enabling the maximum conductor cross section to be suspended in the magnetic gap. All wide range compression driver diaphragms (Fig. 6.15) are coupled to the horn throat via a **phasing plug,** a piece of acoustic plumbing that ensures sound reaches the horn throat from all parts of the diaphragm at the same time, thus avoiding cancellation of high frequencies up to cut-off, between 15 and 20 kHz. The upper frequency limit of a compression driver is subject to the size of the necessary movement gap between the phasing plug and the diaphragm.

Efficiency of a horn loaded speaker is in the order of 40 to 50 percent, compared to 10 or 15 percent for direct radiator cone speakers. This is due to the ideal coupling of driver diaphragm to the air afforded by the exponential horn, which acts as a step-up transformer, or lever.

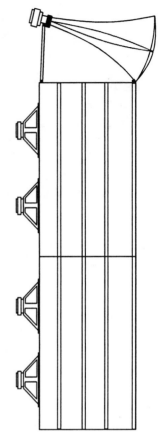

Fig. 6.14 **Traditional theater speaker, side elevation.**

Since large power amplifiers have become universally available, the W-bin is being phased out, first by ported bass enclosures with flared short horns, and lately by ported enclosures of the conventional type, providing compact installation and extended range with lower distortion.

Sub-bass units are also ported boxes, and serve the 20 to 40 Hz octave in stereo installations (Fig. 6.16). However, horns are still used for frequencies above 500 Hz, the crossover frequency being chosen to keep voice frequencies predominantly with one speaker. The ten octave bands of the audio spectrum are shown in the table in Chapter 15.

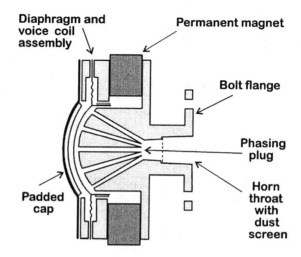

Fig. 6.15 Cross section of a compression driver. Gaskets seal the diaphragm and cap.

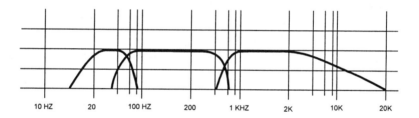

Fig. 6.16 Components of Motion Picture Theater acoustic response, with emphasis on the sub-bass function as an extension, not a boost.

IMPORTANCE OF THEATER SPEAKER DISPERSION ANGLE

High frequency horn speakers for non-cinema use sometimes cross over at 750 or even 1200 Hz to increase HF driver power handling capability by limiting the bandwidth, but this method can't be used for cinema because the high frequency horn is required to cover most of the dialog spectrum, to make use of its single point origin. Attempts to use smaller horns for cinema sound, even though they might be rated down to 500 Hz, have failed for the same reason; the coverage right to the back of the auditorium just doesn't happen over the whole dialog spectrum.

A similar comment applies to the use of a single horn to handle all the mid and high frequencies. Dialog divided between a mid-range and a high frequency horn works poorly in a large auditorium. Propagation, phase integrity, and clarity all suffer if the spectrum is split up.

An ideal speaker propagation characteristic is a necessity for large theater reproduction of drama dialog, but it would have limited success were it not for its combination with **signal processing** of the pre-recorded program, which was introduced in Chapter 5, and is comprehensively dealt with in Chapter 8. Together, they make possible today's high standards of motion pictures and theatrical dialog.

RECENT DEVELOPMENTS

Constant Directivity HF horns are a later development, eliminating former dispersion problems in a smooth coverage of constant level and frequency response from one side to the other, but most require active equalization. This is not a problem, because bi-amping with active crossovers is replacing passive crossover networks, improving phase distortion and frequency response, and a lot of signal processing is now performed in stereo cinemas prior to amplification, including decoding of analog soundtrack formats and digital soundtrack reading, plus a graphic equalizer on each channel to set up the auditorium response (see Theater and Auditorium Calibration, page 127).

Amplifier power for cinema reproduction has risen over the years, as bass speaker efficiency has fallen. In the 1930s, efficient horn speakers used 15 or 20 watt amplifiers. As technology advanced, 50 watts was used with horn speakers in early stereophonic systems using multi-channel magnetic stripe on the film. The extra power permitted wide dynamic range and active equalization.

With increasing dynamic range available from SVA (stereo variable area) optical sound-tracks with advanced noise reduction formats, and optical digital soundtracks, amplifier powers of 500 watts per channel are now common. A theater may now have anything up to 7,000 watts of amplifier power, inviting the obvious question: "When are we going to use all that power?"

The answer is that we never will, but it is available to handle program peaks on individual channels whenever they occur, and guarantees that we will never hear distortion due to clipping, or damage the speakers with square wave or other distortion products.

The high frequency horns are very efficient, and most will not handle peaks over 200 watts, but the single type of amplifier is valuable for its interchangeability in the event of a breakdown.

Stereo Theater Sound System Line-up

H.F. Left	500 watts	L.F. Left	500 watts
H.F. Inner left	500 watts	L.F. Inner left	500 watts
H.F. Centre	500 watts	L.F. Centre	500 watts
H.F. Inner right	500 watts	L.F. Inner right	500 watts
H.F. Right	500 watts	L.F. Right	500 watts
Surround left	500 watts	Surround right	500 watts
Sub-bass 1	500 watts	Sub-bass 2	500 watts

Plus a spare 500 + 500 watt amplifier.

CINEMA HIGH FREQUENCY ROLL-OFF

The frequency range of a modern picture theater is very different to that of a few years ago. An acoustic reproduction standard broadly called Academy Roll-off was used until the advent of stereo variable area to monitor and exhibit soundtracks around the world (see Fig. 6.17). In high fidelity terms, it was fairly restricted, being around 13 dB down at 10 kHz. But it was quieter than 78 RPM shellac phonograph records, and movies sounded good.

The purpose of the standard was two-fold. It ensured that good film product would sound acceptable in most theaters with imperfect acoustics and basic equipment, and the limited high frequency response of reproducers ensured that noise would be attenuated to tolerable levels; noise that was due in part to film emulsion grain particles, film scratches, and

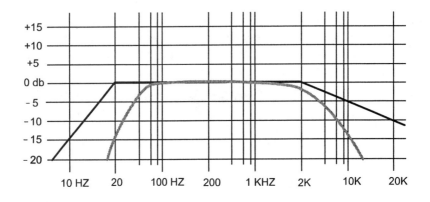

Fig. 6.17 Academy roll-off compared to stereo acoustic characteristic.

photo-electric cell hiss. The technology has improved with the advent of solar (photovoltaic) cells for optical pick-up. They are quiet and have a flat frequency response if amplified as a current source. This is of particular value since the scanning beam height was halved with the advent of stereo to achieve extended response.

There was no point in setting a standard so high that thirty percent of world cinemas could not meet it. Restricting the response of reproducers was to mean use of the same limitations in the mixing theatre, so to compensate, extensive dialog processing was performed to produce clear, unambiguous diction.

Unlike magnetic recording, phonograph records, and FM broadcast transmissions, there is no equalization characteristic applied to optical film recording and playback, but a high frequency recording boost called slit-loss equalization is applied to compensate for the size of the scanning beam when optical sound prints are reproduced. However, use of the restricted monitor imposes its own inverse characteristic when sound-tracks are balanced and mixed. The established formula for dialog equalization and processing remains basically the same even though reproduction has considerably improved, because it suits the acoustic limitations of large theaters.

Motion picture sound evolved over several decades instead of taking a leap into high fidelity, partly because of the large amount of hardware and existing program material that had to work for engineers, film producers, and exhibitors, but also because of the impossibility of upgrading all the exhibition theaters in a short space of time. Product compatibility was the first consideration, so stereophonic optical soundtracks were introduced in the late 1970s /early 80s in a form that was compatible with existing monaural equipment. Later, noise reduction principles were extended to the limit and could no longer be reproduced on mono sound systems. Now, both analog and digital soundtracks in several forms are produced, requiring add-on reproducers and decoders to be fitted to existing equipment.

Six-track magnetic sound with noise reduction on 70 mm film is still used for major motion pictures, but the cost per print is several times that of 35 mm film with optical soundtracks.

Movie sound now enjoys high fidelity status, but there's a second, less obvious reason for the 3 dB per octave roll-off above 2 kHz shown in the auditorium acoustic response of Figs. 6.16 and 6.17, which is the stereo upgrade compatible response characteristic to which theaters were converted.

AUDITORIUM RESPONSE LIMITATIONS

All sizeable auditoriums have one thing in common; a variety of different length multiple reflecting paths, and these set up conditions that make short duration, or transient sounds, perform differently to sustained, or steady state music tones or test signals, by reason of high frequency cancellation when the signal continues beyond the natural reverberation time. For just as reverberation falls away, or decays, in a period defined as when sound level drops to 75 percent of the original, so too, continuing sound becomes further sustained in a similar period by the build-up of reverberation.

The result of this factor is that the apparent frequency response of dialog transients, the parts of speech that carry intelligibility, or understandable data, is different to the response of continuous signals. It is a point to be considered when calibrating a theater, since a flat acoustic response would sound too bright on normal dialog, and would attack the listeners' senses in a manner inducing listener fatigue in a fairly short time. Music and sound effects are less affected, because their content is different; their tones are for the most part longer and more subtle than speech, and they don't carry the same type of essential message. The senses can't be tricked with dialog, the listener knows when it is unnatural. A valid comparison would be the color balance of a photograph, where the skin tone of faces has a strong recognition factor compared to general scenery, or objects whose color is non specific.

Drama voice reproduction has to be very good to be believable in areas where every word has to be understood, and mixing theater calibration is the critical factor that centralizes the tonal balance of programs that can be replayed successfully in cinemas where the acoustic alignment is out of the producer's hands.

MONITOR SPEAKERS

A monitor speaker's purpose is to give the operator a true reproduction of the program, centrally representative of the speakers and the acoustics where the program will be heard after distribution.

The speaker chosen for monitor use doesn't necessarily have to have a flat frequency response. That can be fixed together with listening room errors with a calibrated graphic equalizer, but it should be free from harmonic and phase distortion, and in particular from response peaks which would give a false impression of the program's brightness, bass content, or intelligibility.

Response peaks can be quite narrow, and they don't always show up on a spectrum analyzer, nor can they be corrected with a equalizer, the only choice being to avoid using speakers with this type of error. Similarly, the dynamic performance of different speakers is a factor, as a monitor must remain true up to its maximum level of use. Most music is monitored at high level at the mixdown, and the speaker and the room need to be free of effects that would alter the perceived balance at the highest levels.

A monitor must be made true to the parameters of the program's ultimate presentation; a **central representation** of the way others will hear it when reproduced later on a variety of systems. It must also be set up in a listening room where the acoustic perspective centrally represents the **ideal** of the final listening situation. Motion picture soundtracks are mixed in big rooms representative of cinemas. The same perspective is not really achievable in small rooms.

Harmonic distortion and response peaks are particularly hazardous as they come under the category of "synthetic highs"—apparent brightness that is not really there. For the speaker to be indulging in fantasy is not a good trait in a monitor that's supposed to tell the truth, and it will change at different levels, which is one reason why it is the practice to calibrate and use monitors at set sound pressure levels.

Response dips, or troughs, are also important, but not as damaging as peaks, especially if they occupy a relatively narrow slice of the spectrum. Balancing to an over-bright monitor will produce the inverse; a dull program. Dull monitor; bright program. Bassy monitor; thin program, and so on.

A mix may sound great in the studio, but when replayed elsewhere it can be seriously out of balance. It is important that all the electronics in a monitor chain has high electrical specifications, but the speaker and the room are also parts of the monitor chain and more subject to error. Together they can be out by many decibels, so calibration with spectrum analyser and graphic equalizer is a necessary part of setting up a monitor system.

LIMITED MONITORS

Many studios, especially in the music industry, balance their product on wide range monitors, some at a level representing the live performance, and then double check the mix on a purpose designed monitor of limited response to test the mix at normal listening levels as it will be heard in a high percentage of cases.

Ideal limited monitor speakers have two things in common; they are chosen to be uncompromisingly flat within their limited passband and free from harmonic errors. They sound disturbingly natural on that intensely recognizable sound, the human voice. Use of a wide range monitor for absolutely everything can result in apparently different levels when reproduced by many final listeners on limited range equipment. This is never more evident than in live television broadcasts where field reports are intercut with studio sound, and with less mid frequency presence, balanced for level alone on the studio monitor, they can appear to be lower in level than studio sound when the bass region is attenuated by the small speakers in the majority of TV receivers.

Not all program producers enjoy the luxury of time to check their product on a limited monitor before presenting it.

AUDITORIUM AND MONITOR CALIBRATION

The calibration procedure is not nearly as complicated as it appears at first. It takes relatively little time to master, and especially important is the technique of making several quick alignment runs using small adjustments each time, because each alteration affects adjacent parts of the audio spectrum. The technique outlined here will get consistent results in cinemas, AV theaters, public auditoriums, and home theater and stereo installations.

Both fixed and portable sound systems benefit from acoustic calibration, a straightforward procedure requiring permanent installation of a 10 or 31 band graphic equalizer in each channel.

The test set up (shown in Fig. 6.18) consists of a pink noise generator, and a spectrum analyser with omnidirectional microphone. If the analyzer has filters for simulating A, B, or C weighting curves, they should be switched out so that flat frequency response is displayed. Some analyzers include a pink noise source, otherwise a separate generator is needed. The one I use is in a plastic box about four inches square, and is run by two 9 volt batteries. Preliminary calibration of the pink noise generator and the spectrum analyzer is performed by sending pink noise directly into the analyzer's line input, and noting that the response is flat.

Pink Noise is a broad spectrum signal containing all audio frequencies, and has equal power per unit bandwidth; in other words, each octave of a pink noise signal across the spectrum has equal power. Each of the thirty-one third octave bands of the analyser will read the same level on a flat response pink noise signal, which is derived by equalization from White Noise, a more familiar term. White noise is the signal produced by electronic noise generators of various types; it sounds like a "hiss" and has a rising characteristic of 3 dB per octave towards the high frequency end, which makes it unsuitable for calibration without modification.

Sound systems are made up of two main parts:

1. The 'A' Chain: Sound sources, mixer, source selector or control unit, and main gain control up to line level. In a cinema, this includes the stereo processor.
2. The 'B' Chain: Monitor gain control if it is additional to the main gain control, auditorium equalizers, speaker processors or equalizers specific to the speakers used, power amplifiers, and speakers. The B Chain really includes the auditorium, because although it is not an

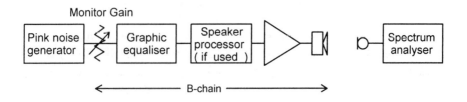

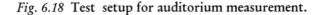

Fig. 6.18 Test setup for auditorium measurement.

electro-mechanical device, that is where the response will be measured, and it's the component in the chain which probably requires most correction.

Calibration proceeds as follows:

1. Install a third octave graphic equalizer in each channel just before the power amplifier. If there are non linear devices like limiters or compressors in the system, they should should be switched out during calibration, but speaker processors or equalizers which are part of the speaker system should be included immediately before the power amplifiers, with speaker line sensing limiters turned off for the calibration.
2. Set up a calibrated omnidirectional microphone in the auditorium, about two thirds back, slightly off center, angled up to the stage speakers. Fit a random incidence corrector to the microphone if one is available. The height should be about 4 feet, to avoid interaction with chair backs. If surround or ambient channels are also to be corrected, the forward-facing microphone orientation should not be changed.
3. Send pink noise into the system, one channel at a time, at a line-level point before the graphic equalizer, and measure each channel at a maximum auditorium level of 85 dB. Choose the level with care, as continuous pink noise can damage speakers. If speaker damage is suspected, reduce the SPL (sound pressure level) to 80 dB, and/or allow a five minute cool-down period, and measure in bursts of thirty seconds, separated by thirty second rest periods. The amount of power needed at the speakers to achieve pink noise output at a usable level depends on the type of speaker and the size of the auditorium. Choose the acoustic response characteristic you intend to use; flat for general use and home theater, and a 3 dB per octave fall above 2 kHz for cinema, or programs with prominent dialog.
4. Ambient noise can make the measurement difficult to read even if sufficient level is used; generally the bass spectrum is most affected. If the air conditioning can not be turned off or low frequency noises interfere with the measurement, observe the noise spectrum with the pink noise off, and compensate for it when making corrections by adding half the indicated noise level to your response reading at each frequency band affected. Switch the pink noise source off and on again several times, and the true response reading at each frequency band will be revealed.

5. Observe that the series of peaks and troughs in the bass, shown in Fig. 6.19, are probably due to standing waves caused by reflections generating interference patterns. Ironing out these wrinkles completely will only make them worse at some other microphone position in the auditorium. It is a good idea to average low frequency error without trying to straighten out every cyclic change. Serious treatment would require acoustic absorber panels on the back wall of the auditorium to control reflections. A dip at 500 Hz may be evident on some systems with passive crossover networks. It will correct reasonably well. A serious dip may indicate that a driver is out of phase. Adjacent analyser and equalizer bands interact, and it is better to go over the spectrum two or three times with partial corrections, rather than try to make all the corrections in one pass. If the corrections don't work out and the procedure gets bogged down, reset all graphic equalizer controls to zero and start again; it will save time. When equalizing, generally, cut before boosting, and be aware that the average maximum level should remain the same at average equalizer unity gain.

6. Don't swing adjacent equalizer controls to opposite extremes, this introduces serious phase shifts, and avoid applying too much boost or cut at any one frequency.

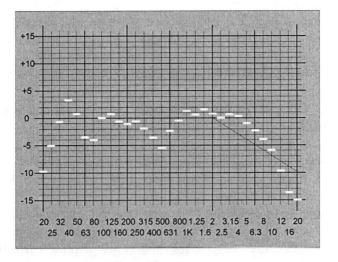

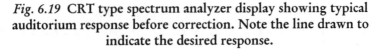

Fig. 6.19 CRT type spectrum analyzer display showing typical auditorium response before correction. Note the line drawn to indicate the desired response.

7. After each channel is reset to unity gain following calibration, reset the gain to each channel and send pink noise into all channels simultaneously. Observe that the analyser display is no longer flat, but don't make any changes. The combined response will look bad on the analyzer because all channels are reproducing an identical signal via multiple length direct and reflecting paths, which are generating sum and difference interference which won't be there when normal program is on the speakers because the signals will be different. But this is a useful test that will pick up an out of phase speaker.

8. Look at each channel one after the other, and re-equalize any that could use improvement, but remember that the general shape of the response curve is far more important than small peaks and troughs that can't be fixed without incurring some other penalty.

9. Finally, listen to a variety of known quality material. Do not use second-rate program for this purpose. At this point, given the variations in bass and apparent brightness that auditoriums of different reflective qualities can produce, it is quite in order to make small subjective alterations to the overall bass and treble balance in an exhibition theater, but inadvisable in a studio monitor situation where a true response characteristic is critical.

Some auditorium equalizers make provision for this by adding bass and treble controls, otherwise it will be necessary to make minor changes to a few of the controls. Fit a security cover to each graphic equalizer. The subjective program listening test will pick up gross measurement errors, so at this point, if things just don't sound right, recheck the procedure and recalibrate the auditorium.

When one such calibration procedure has been experienced, it will be realized that each spectrum adjustment run is just preparation for the next, and that while some auditorium setups proceed without a hitch, others can be very reluctant to come together, but once set, they don't drift; barring major building alterations, decor changes, or the introduction of a lot more furniture.

The audience will not be there for the calibration, and if serious changes are expected when the house fills up with people, a high frequency lift of one or two decibels can be made. Anticipating a high frequency loss is tricky; mixers do it when balancing programs, for there is a brightness threshold that makes acceptable a program that is slightly over-bright, but sound that is dull, or too far below the threshold suffers an unacceptable loss of intelligibility.

CALIBRATE AN AUDITORIUM FOR AN ESTIMATED FULL AUDIENCE.

Anticipating an audience makes a bit of extra brightness acceptable, as the auditorium will be empty when it is calibrated.

DO NOT ARBITRARILY ADD BRIGHTNESS TO A MONITOR. IT WILL DULL THE BALANCE OF YOUR WORK.

Monitors need to be regarded differently to auditorium speakers. They work inversely; bright monitor, dull mix; bassy monitor, thin mix, and so on. There won't be a full audience when subjective judgments on current product are being made later, either during mixing or previewing.

CALIBRATE A MONITOR ACCORDING TO SPECIFICATION.

Bear in mind that large amounts of response correction will have the same effect on the speaker's power handling specification as any abnormal equalization made to the program. Boosting an appreciable part of the spectrum reduces the power rating of the speaker by the amount of the boost.

SPEAKER RESPONSE MEASUREMENTS

It is often intended to measure and/or calibrate the response of an actual speaker, as differentiated from the auditorium-plus-speaker combination, and this is done for two reasons. First, it may be of interest to know the response of the speaker, because many domestic speaker manufacturers don't publish the on-axis response as it only tells part of the story. If they do, it's an aim response, achieved while developing the prototype, and it doesn't include a couple of peaks here and there, and won't include production and aging variations. These peaks and troughs are not serious faults in a speaker intended for general or domestic use, because a reflective domestic listening room covers up a multitude of small errors by averaging the defects, and normal degrees of deviation from the ideal will seldom be noticed; many of them are minor compared to the errors produced by most rooms.

Second, it may be intended to set up a graphic equalizer to represent a lost or otherwise missing speaker processor. Several professional and some domestic speaker brands use an outboard line-level equalizer, which may also contain a speaker-line sensing limiter, or an active crossover. The speaker process equalizer is a designed part of the speaker, made for insertion before the power amplifier to boost the bass among other things so that a smaller speaker box can be used, and/or to produce a flat acoustic response. Some of them have large amounts of frequency correction and the speakers cannot be used without them. A 10 band, or preferably a 31 band, graphic equalizer will do the job, probably not quite as well as an original dedicated equalizer, but better than nothing. The response curve of a calibrated graphic equalizer is not likely to be exactly the same as an original process equalizer, so an estimate of reduced power handling capability should be made for any speaker so equalized.

The measurement should be made in a furnished and carpeted room, with the measuring microphone on front axis, 3 to 6 feet or 1 to 2 meters away. Raise any speakers less than half a meter high so that the microphone axis is one third from the top of the speaker, or central, depending on the driver deployment. The distance should be short to optimize the direct-sound to room-reflection ratio, but sufficient to average the combined output of all the speaker's drivers and ports.

Speakers with drivers and/or ports at the sides or back should be placed in the manufacturer's recommended position in relation to back and side walls of a room of brick or well supported plaster-board so that the **short path** reflected outputs of all drivers and ports mix as they would in normal use. Professional speakers also need a backing wall as depicted in Fig 6.8. It is not possible to make one rule covering all circumstances where side or rear output speakers are concerned, but the most important part of the setup for this type of speaker is to have a reflective wall or corner behind it, and set the microphone far enough back to include a suitable proportion of the reflected sound.

Speakers should be measured one at a time, as in auditorium calibration. The combined output of two speakers reproducing the same program won't be of any value to the measurement as it produces interference patterns which invalidate the measurement. A speaker measured on axis at close range should be calibrated for a flat acoustic response, not a rolled off HF as in some auditorium characteristics. The main reason for such a roll-off is given on page 125, and is connected with variable length multiple path reflections. These are not significant in short-path measurements, and they should be discounted.

If you find that a speaker exhibits a mess of peaks and troughs, making it impossible to see a clear indication of its general balance, turn it around to face the wall, about eighteen inches away, and look again. The response will flatten out, displaying the general balance of highs, mids and low frequencies, because it is one of those useful facts that short multiple reflection paths sum to iron out minor deviations in response. If the measurement is part of a calibration series, make some adjustments, then turn it around again to face forward. The analyzer display will now make more sense, and calibration can proceed.

The maximum amount of boost that should be given a speaker, above its most sensitive frequency band, is about 10 dB. In the case of speakers for which the original process equalizer is not available, reference should be made to any known information about the speaker's performance to determine the natural end frequencies so that the permissible extent of equalization can be determined.

Commercially made speakers have undergone hundreds of hours development; it is unlikely that their performance can be reliably enhanced by anything simple that can be done to them later without incurring a penalty of some kind. General modifications to commercial equipment are in the same category; without equivalent research facilities and construction of further prototypes, it can be taken for granted that a particular unit has been developed as far as it can be, without incurring negative trade-offs or capitalizing it beyond its value.

AUDITORIUM AND MONITOR LEVELS

Sound reproduction levels vary enormously with different types of program and venues. The 85 dB Acoustic calibration level used in auditorium equalization was originally equivalent to 50 percent modulation in movie soundtracks; about the level where most of the dialog is located, because that allowed headroom for shouts and loud effects, but recently the dynamic range of extended analog and digital tracks has increased considerably. Soundtrack mixes are monitored at the same level as cinema presentations, and other media like the music industry often monitor at high levels representing the sound pressure level of the original performance, the purpose being to hear the right perspective when balancing and equalizing.

The frequency response of our hearing changes with loudness. At high level, all frequencies are perceived to be almost equally loud, but as the level falls, the high and low frequency ends of the audible spectrum appear lower than the middle band of frequencies between 300 and 6,000 Hz. The effect of this, when listening to music at home listening levels, for example, is to make a normal bass content almost disappear if the volume is lowered 10 dB.

Loudness controls are volume control circuits with equalization that tracks along with the attenuator, usually boosting bass only as volume is reduced. Unfortunately, lack of world standardization limits the usefulness of such devices, which are used in some consumer products, coupled with the need for correct input level to place the equalizer in the correct volume range for it to work properly.

Since they are an imprecise tool, therefore, loudness controls are not used by mixers. Instead, they monitor their programs at consistent levels, often of their own choice, depending on the type of program. This ensures that the audio frequency spectrum remains in the same perspective from one end of a mix to the other, and from job to job. Speakers and auditoriums also change their characteristics if pushed beyond their optimum **sound pressure levels**, so like frequency response measurements on tape recorders and the like, calibration levels are made below maximum level to guarantee that the equipment and the acoustic environment is not adding overtones and non-linearities of its own. Overtones are extra unrelated frequencies, similar to adjacent strings on a piano that vibrate in sympathy because they are close in pitch.

WHY DO COMMERCIALS SOUND LOUDER ?

Movies set most dialog at 50 percent modulation, or minus 6 dB on the VU meter scale (Fig. 8.3). This leaves room for shouts, close-ups and loud effects. TV drama sets dialog somewhat higher due to the ambient noise of listening conditions. But an advertisement is a short presentation which seldom has time for dramatic perspective, and there is less requirement for it in light of the advertiser's pressing need to get the message out and catch the listener's attention. So if people complain that the commercials are louder than the programs, they're right. However, some TV channels compensate and run commercials at lower levels.

HEADPHONES

Monitoring a mix on headphones presents a totally different perspective of the program content, compared to monitoring on speakers with added room acoustics. While they are great for personal listening and help concentration when checking for defects in a recording, headphones can not be used to monitor a program destined for normal reproduction because the resulting balance will be quite wrong, and instruments, effects, and the stereo image will be found to have disappeared when the mix is replayed later on speakers.

Chapter 7.

Tape Recorders and
Monitor Selectors

This chapter deals with **Tape Recorders, Monitor Selectors,** and the relatively unrelated subject of **Schematic** and **Block Diagrams,** but the purpose of this will be made clear.

There are two things that are faced sooner or later by everyone in a technical career or pursuit of interest. The first is the need to acquire basic drawing skills, so that at least a road map of the current project can be made, to inform others, and to further one's own understanding. Whereas a schematic diagram contains all circuit details, a block diagram is nothing more than a set of function boxes connected by lines. Drawing a block diagram of something I am working on helps me think; I can see it in front of me, so I can't forget the part I've already done, and I can see at a glance if an idea is complete, or if it needs further development. Even a rough pencil sketch will do for a start.

The other thing is the need to have a routine method of getting to know a fairly complex piece of equipment. We should never be intimidated by unfamiliar circuitry, since audio or control engineering is just a sequence of simple principles. When broken down into its several functions, each one can then be examined in more detail. But taken as a whole, most systems or equipment can be viewed as nothing more than a series of labelled boxes joined by lines. That's the block diagram.

About the only guidelines for system diagrams is that the operation should, in general, flow from left to right, like writing, with arrows added if it will clarify the drawing or if function flow has to reverse direction. Also, that that the building blocks should be labelled, or coded with symbols as in Appendix B at the back of this book. This makes diagram communication universal, so that anyone can read your drawings without requiring explanation, as in the following block diagram of a tape recorder (Fig. 7.1).

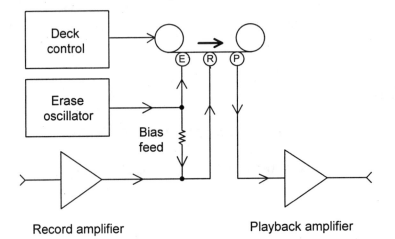

Fig. 7.1 **Three-head tape recorder block diagram.**

MAGNETIC RECORDERS

The general arrangement of a tape recording or playback head is shown in Fig. 7.2. Coils, arranged in opposition to cancel external magnetic interference fields, are wound on pole pieces of various shapes that form the magnetic core.

The two halves of the head core are separated by shims of non-magnetic material to form gaps across which the recording field appears; the front gap transferring signal fields to the tape oxide layer, and the back gap also separating the pair of opposing pole pieces so that the head will be balanced against external magnetic fields.

A magnetic shield encloses the play-back head, because a high gain amplifier follows it, and the small signal voltages have to be isolated from external noise fields. The tape wraps around the head at a small angle consistent with low wear and good contact pressure at the gap.

Bias is supplied by the erase oscillator because the erase and record bias frequencies must be locked together to avoid heterodyne generation (sum and difference frequencies), which can cause low frequency noise build-up on successive recordings. The high frequency bias, which may be anything from 40 and 120 kHz in a real time recorder, actually does record, but at a low level.

Recording takes place on the **trailing edge** of the head gap as the tape leaves the bias field, and the highest frequency is determined partly by

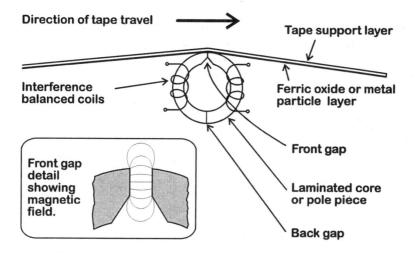

Fig. 7.2 Magnetic tape in contact with record or replay head,
showing construction.

tape transport speed, but also by the bias frequency and level, the oxide type, and head inductance and capacity.

The relationship of head gap dimension to the wavelength of the highest intended frequency is the key factor in playback response, as replay takes place **across the gap**, which is smaller in playback than in record heads. Record and erase heads have lower inductance than playback heads.

The recording field permeates the oxide layer as the tape moves past, producing a series of magnetized zones that will generate corresponding currents in the coils when the tape makes the replay pass.

It does not require much signal to magnetize a tape; line level will do it, but because of **hysteresis**, the magnetic "elasticity" of materials, the program signal by itself produces a poor transfer characteristic.

Magnetic recording only achieved high quality with the development of high frequency bias in the 1940s.

Hysteresis is the property of all magnetizable materials whereby it takes more field strength to magnetize them than they will retain (Fig 7.3a), and it makes necessary the application of a high frequency bias voltage, added to the signal, which cycles the oxide through its positive and negative magnetization states many times for each portion of the signal waveform, leaving the tape magnetically neutral, as would a demagnetizer; or **degausser**, except for the magnetization due to the signal (Fig. 7.3b). The bias frequency chosen depends on the tape speed and the bandwidth to be recorded; the amount of bias being chosen to maximize high frequency response for the particular oxide in use. The bias voltage, as well as the frequency, is several times the audio signal voltage, which envelopes the bias (Fig. 7.4).

Insufficient bias will result in distortion; too much bias reduces signal level on the tape, especially the high frequencies. Metal particle tapes require more bias than ferric oxide coatings.

Adjusting the bias involves advancing the bias strength until the high frequency response of the simultaneous play-back just commences to fall. On a two head machine, where playback is taken from the record head, it is necessary to record an incremental bias **test series**, enabling the optimum setting to be chosen when the recording is played back later.

Low distortion bias waveform is required for good recorder performance. Bias harmonics and magnetized heads will result in recorded noise as well as distortion. The same general theory applies to magnetic recorders of all types, including video recorders, but the design and materials of more advanced heads makes them relatively immune to accidental magnetization.

Magnetic tape is traditionally made with a coating of ferric oxide in the form of needle shaped crystals in a lacquer base, which includes silicone as a lubricant for the heads. Other materials are used, sometimes as a thin layer on top of the ferric oxide. The oxide is magnetically grain-oriented on the tape during manufacture to optimize it for the direction of recording. Ferric oxide is prepared from a solution of iron sulfate to control particle size and purity, and is then ground into a lacquer base in a continuous multi-stage ball mill to produce

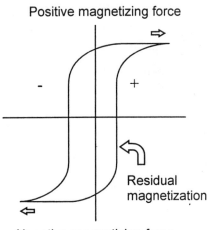

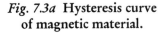

Positive magnetizing force

Residual magnetization

Negative magnetizing force

Fig. 7.3a **Hysteresis curve of magnetic material.**

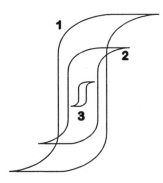

Fig. 7.3b **HF bias decay of an oxide particle.**

the required fine surface. The tape finally goes through a calendering process, where it is pressed between hot rollers, as in paper or textile manufacture, to finish the surface.

Technique and materials have advanced considerably since high frequency bias was developed, for without improvement over early recording materials and heads, much of the video recording practiced in domestic and professional fields would be commercially non-viable.

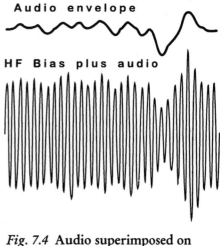

Fig. 7.4 **Audio superimposed on HF bias.**

RECORDER FEATURES AND TYPES

The complexity of any equipment depends on its purpose, and the type of market it is made for. Domestic recorders can be loaded with useful features, like automatic level control, track to track transfer, multiple speeds, and so on; a studio in a box, complete with pre-amplifiers and microphones.

Studio recorders are heavy on control systems and special drives with servo-motors, some types use **time code** which enables them to search, interlock, or edit to order, but in audio terms, they tend to be straightforward, having an input and an output at line level, with or without a noise reduction system, and a sync-play output which replays off the record head so it will be in sync with other tracks being recorded. They usually have a sophisticated insert editing facility, which switches erase current on and off sequentially, delay-synchronized with the recording bias, so that gaps and overlaps are not left on the tape when a new section is inserted into an existing recording (Insert Recording, page 151).

Some professional portable recorders have servo-capstan drive and a a pilot tone channel, with an add-on or built-in play-back resolver, so that they will record and play back speed-locked to an external pilot tone source, or an internal crystal locked pilot tone generator. Pilot tone heads have two anti-phase tracks to effect mutual rejection with program recording. Pilot tone capability makes it possible to separately record sound with a film or video camera, for later replay in sync. Otherwise,

time code can be recorded on a spare channel. Time code is a digital
address corresponding to each **real time** fraction of a second.

SIMPLIFYING COMPLICATED CONCEPTS

Not long ago, the technician in charge of a government installation
I was servicing asked, "How does television actually work?" I started
explaining the carrier frequency bands reserved for transmission, the
bandwidth needed for video signals, and would have carried on for
five minutes, but he pulled me up: "No, how does the picture get on
the screen?"

In no time at all, an impromptu speech was delivered on how the
spot of light starts at the top corner of the screen, writes an intensity
modulated line of light across to the other side, and flies back to start
the next line under the first, like a typewriter, continuing down until
a picture fills the screen. Then it does it all again, several times a
second.

It isn't complex, really, but in most electronic systems, there's a lot
lot happening at the same time, and if this analysis is applied to even
the most intimidating systems, nobody will ever have to turn and
run when confronted by an expensive looking box covered in lights
and buttons.

Although this chapter introduces recorders so that we can talk about
monitor selectors later, it is a good opportunity to explain the principle of
Analysis by Diagram. The following basic descriptions break down
equipment into their component function blocks. To create anything at
all is surely just a matter of selecting blocks, say, numbers 1, 3, 5, and 27,
and assembling them in the right order:

- A radio/cassette recorder is a receiver with a tape recorder and
  speakers.
- A television set is a receiver with a screen and speakers.
- A video cassette recorder is a receiver with a tape recorder, a timer,
  and an RF modulator.

Simplified descriptions are OK, but a better way is to **Block Diagram**
the equipment, and demonstrate once again that a picture is worth a
thousand words. We've already seen the diagram of the audio recorder on
page 138, but here it is again in more detail (Fig. 7.5). Now we can take
the next step and draw a VCR.

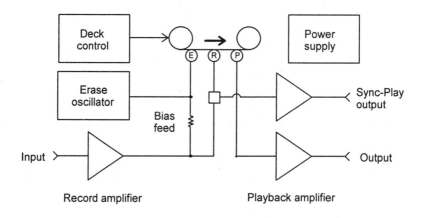

Fig. 7.5 **Three head tape recorder block diagram.**

There's more in it than an audio tape recorder, and that doesn't count the extra control systems. But when a block diagram is drawn, the VCR is revealed as a tape recorder with just more of the same thing (Fig. 7.6).

No details are shown of the control track system, which records a 240 Hz sine wave on one edge of the tape, because it's automatic, and it would spoil the diagram's simplicity. Hi-fi stereo replay off the video head has also been omitted for the same reason, but we know it's there.

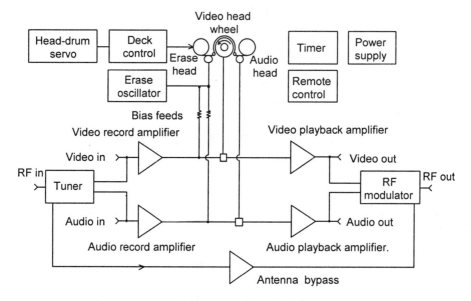

Fig. 7.6 **Video recorder block diagram.**

Just like the engine and transmission management computers in a car, the operator does not have to think about control tracks and servo-motors, they do their job, and have no interface with the world outside their box except the control track the system records on the tape.

What concern us in the context of this book are the program channels. Like the audio tape recorder, there are record/play video and audio heads, and the record and replay amplifiers.

Drawing techniques are not seriously discussed until Chapter 14, but drawing up a system as a block diagram is about the best way to get to understand it, and is an easy exercise, because if it starts to look too busy, you just leave out circuit lines that spoil the clarity of the drawing, provided they are understood to be there; for example: the timer and remote control connections on the VCR.

GENERAL PURPOSE MONITOR SELECTORS

If a **monitor bridge** serves only one source, it can be as simple as a potentiometer and a bridging resistor on a speaker line (Fig. 7.7). In its simplest form, a **monitor selector** can be a rotary switch which sweeps the required sources or monitor points (Fig. 7.8a).

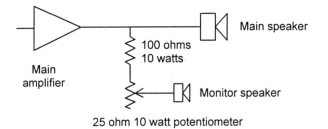

Main speaker

100 ohms
10 watts

Main
amplifier

Monitor speaker

25 ohm 10 watt potentiometer

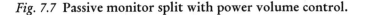

Fig. 7.7 **Passive monitor split with power volume control.**

The system has obvious drawbacks, and is better done with a separate monitor amplifier bridging either the speaker line or the line level source. When this is done, there is an opportunity to sweep other sources at the same level, making a monitor selector suitable for fixed and mobile sound installations. If the switch is a "make-before-break" or shorting type, then bridging resistors must be used at each monitor point to prevent momentary cross-talk across adjacent switch contacts, and in any event at least one bridging resistor is required to avoid loading the circuit to be sampled,

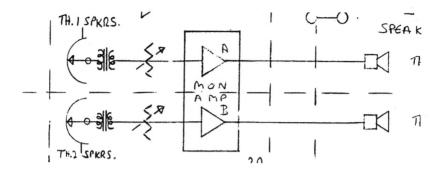

Fig. 7.8a   Rotary monitor selectors in a system draft plan.

which could result in level changes or distortion. If small DC potential differences between sources cause audible clicks as the switch is operated, then a 10 megohm resistor across each switch contact will bleed away the voltage and silence the switch.

A more elaborate monitor selector, suitable for transfer suites or sound systems uses a row of key switches to enable any source to be compared with any other in an A/B comparison, without having to pass through other source selections. The one shown in Fig. 7.8b includes a VU meter. VI, or Volume Indicator, is the name used for an unspecified meter. A VU (volume units) meter is shown with a 3k6 (3.6 kilohms) series resistor which would scale a VU meter correctly to read zero VU at +4 dBm in a 600 ohm line. If a VU meter is included, individual monitor point bridging resistors are not used, as that would spoil the meter calibration, so non-shorting break-before-make switches are chosen.

A compact monitor selector can be built on a two rack unit panel, or even a corner of a speaker patch-bay, but mixers invariably use a **control grouped** panel of more ideal dimensions. The principle of control grouping will be mentioned in more detail in Chapter 14, in the section on

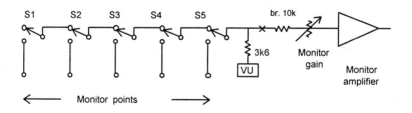

Fig. 7.8b Sequence-switched monitor selector.

panel design, but basically it is a layout protocol that spatially segregates the functions to make a group of switches and their designations more instantly readable, and recognizable to the hand if the eyes are busy elsewhere, or the operator is in semi-darkness. A control grouped panel design is more operator-friendly than a matrix of identically spaced controls, as can be seen when comparing products like infra-red remote controls of different brands.

It is convenient for monitor sources to all come up at the same level. Normally, program sources are at standard line level, but if for any reason they are not, then preset variable attenuators can be included to balance them. For example, a remote speaker line might be included in a group of line-level sources. The usual speaker line voltage, probably higher than line level, would be padded back to come up at the same level. Perhaps it would have a pre-set potentiometer, so that it could be revised as necessary to accommodate different conditions of use.

Monitor Selectors sometimes switch both sides of the line, and proceed to the Monitor Gain Control and Monitor Amplifier via an **audio isolation transformer**. This is done in cases where some or all sources are balanced. Sometimes an unbalanced selector is satisfactory, with or without a transformer, where all sources are in the same rack, are unbalanced, and can use a common neutral. However, there is a certain risk of incurring noise or instability when the neutrals of several sources are linked, and it is an inadvisable practice in installations of any great size or diversity.

The reason is that sources located some distance from each other, where there is a noise potential difference in the common neutral wiring, may interact to introduce noise into the various sources. In addition, power amplifiers, which may operate perfectly as stand-alone units, can oscillate at ultrasonic frequencies if input and output grounds are linked outside the amplifier, due to phase shifts outside the audio band. The grounds may be common inside the amplifier, but the place where they go to chassis is carefully chosen by the manufacturer.

The diagram of an unbalanced monitor selector in Fig. 7.9 shows one method of avoiding noise loops when connecting to unbalanced program sources. (Ref. Chapter 10: Noise Immune Lines).

There is a general principle in audio wiring called **star point grounding**, which recommends that program neutrals should not be linked by more than one path. Rather, system grounds in unbalanced equipment should fan out from a single **star point**. Equipment cases and racks have a network of random noise currents flowing though them which are

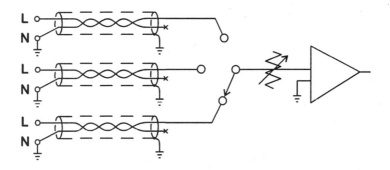

Fig. 7.9 Monitor selector showing floating neutral balance
conductors from the sources, and a grounded amplifier neutral.

caused by leakages and magnetic fields from transformers and nearby
power wiring. These stray currents can cause hum in a system where they
intercept program neutral lines, so when planning a system, avoid having
leakage currents and program neutrals sharing the same conductors, even
for short distances. The diagram of Fig. 7.9 is a compromise which is
usually better than the alternative of linking the neutrals of the sources.

Monitor selectors are in many cases a last minute addition to a system.
They are often not included by specification writers, instructions can be
sketchy, and leave much for the installer to interpret. But it is obvious
that some systems would be unmanageable without some type of monitor
control, which can be very simple, involving only a gain control and a
rotary switch in the corner of a speaker patchbay, a stick-on lettered scale
and designation list under a piece of clear acrylic sheet, plus a handful of
10k resistors. An engraver's layout is shown in Chapter 14.

If monitor selectors are combined with panel that has a separate
function, then it should be a panel that is an integral part of the system,
not an equipment module like an amplifier or a sound source that might
have to be removed for service. The monitor gain control should be
included at that position for convenience. It would also be inappropriate
to mount the monitor selector on a microphone patchbay, for example,
where the levels are very different.

As discussed in Chapter 3, power amplifiers are designed to have
sufficient gain to bring line level up to their full output, so the actual gain
of amplifiers is usually based on the maximum undistorted output. But in
monitor applications, there has to be enough gain to replace the loss of
the bridging resistor, plus the loss held by the monitor volume control,
which usually is positioned at 12 o'clock for normal monitoring levels.

This can be served by the available amplifier gain if full output is not required, or it can be augmented by an external recovery amplifier. An alternative is to use a bridging transformer, which does the job relatively loss free, but these are not readily available items, and have to be well shielded. Chapter 10 discusses distortion and frequency response problems if microphone transformers are used above their design level.

INTRODUCTION TO STUDIO MONITOR SELECTORS

A brief summary of existing sound recording methods follows. It will be noted that the magnetic recorder is the only one which fills all the requirements of the three-mode monitor selector, as described in the following pages.

DEVICE	MANUAL EDITING	ERASE & RE-USE	INSERT RECORDING
Cylinder/Disc and stylus	No	No	No
Motion picture optical film (Analog and digital)	Yes	No	No
Magnetic wire recording	No	Yes	No
Philips Miller system (Engraving on black coated tape)	Yes	No	No
Magnetic tape and film (analog and digital)	Yes	Yes	Yes
Thermoplastic recording	Yes	Yes	No
Laser disc and CD	No	No / Yes	No
Solid state recording methods.	No	Yes	Not yet

MONITOR SELECTORS FOR STUDIO MIXERS

Monitoring a mix, whether recording or updating, calls for a more elaborate and operator-friendly switching arrangement for the monitor points. A magnetic recorder is an inherent part of the studio mixing system, and while the level meter on a studio mixer is able to continuously monitor the mixer output lines, the monitor speaker often needs to tell the operator what's happening elsewhere in the system at the same time, like the recorder outputs. The level meters may be switchable to other lines, but this function is usually independent of the speaker monitor selector.

The mixer and recorder are inseparable parts of a studio system, the principle of the specialised monitor selector which accesses them is detailed in Fig. 7.10. Studio monitor selectors are very much concerned with insert recording, or pick-ups in soundtrack mixing and music recording, because they play a vital part in matching levels and sound quality whenever a new recording replaces an earlier section on one or all tracks of a tape.

The studio recorder edits, erases, and insert records; that is, it can assemble-edit a track by making "drop-edits," or "cutting in" on an existing recording with a new one, and it can insert a section of new recording into an existing one, without leaving any gaps.

A studio recorder is a unity gain device which operates at line level. Each independent channel has an input, a simultaneous replay output, and a sync. play output off the record head that is mute during recording. "Replay" and "playback" mean essentially the same thing. Both terms are used in this chapter, primarily so that repetition is reduced, and whole paragraphs do not degenerate into gobbledegook. In computer terminology, **playback** and **record** would be **read** and **write**.

A mixer feeds the studio recorder, and the monitor selector addresses the recorder inputs, replay outputs, and sync. play outputs, which always use separate playback preamplifiers because, coming from the record head, recovery equalization has to include special correction for the wide-gap high frequency loss.

The recorder inputs and simultaneous playback outputs are accessible to the monitor during both recording and replay. Sync. play is available only in play-back, or non-record mode. The purpose of this degree of monitor access is to enable the operator to play back one or more completed tracks in sync. with others being recorded, to assist in balancing and timing the mix.

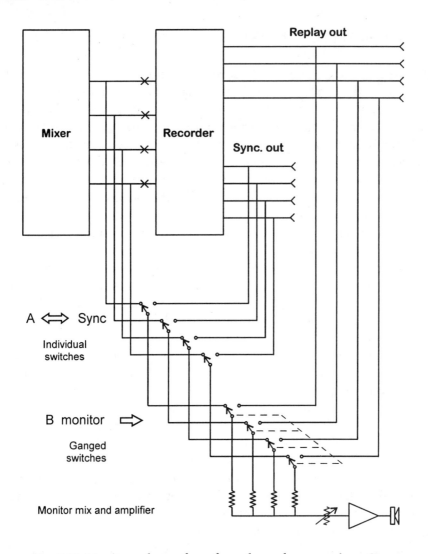

Fig. 7.10 Monitor selector for a four channel mono mixer. Stereo selectors or extra channels duplicate the same circuit. Relays do the work of switches where the number of channels exceeds four, and open microphone situations include a mute interlock on A monitor position unless Sync. Play is selected.

In a live studio recording situation there will be a similar foldback selector which sends one or more tracks to the foldback speakers during recording, to give performers a timing mix of previously recorded tracks to work to. This emphasizes the essential character of sync. replay off the record heads; the mixer's monitor and performers' cues have to be synchronized with the material being recorded, not a tenth of a second late, as would be the case if the playback monitor was driven from the separate playback heads.

An important monitor function is to be able to A/B the recording before an edit, so that recorder input and output levels can be verified before pick-up recording is initiated. Other uses for B monitor are verifying that line-up tones are simultaneously recorded on all tracks at the start of a tape, and checking the quality of the recording with a single switch at any time during its progress.

While recorder input and sync. play are individually accessible, simultaneous play-back, or B monitor, only accesses all tracks together, in order to present all tracks in sync, rather than have some presented late by the delay of the replay head. All monitor points are aligned to present the same level, so that identical recorded levels can be set up when an electronic edit is made.

DROP-EDITS AND INSERT RECORDERS

To enhance the concept of a monitor selector like the one just detailed, a description of the **electronic edit** function of a magnetic recorder follows, and it should be noted that there are several ways it can be done.

A drop-edit can be made by switching on the recording bias and erase current simultaneously, on cue. At that moment, an existing recording running in replay mode commences replacement with a new recording. Doing it this way has two effects:

First, the original and replacement recordings overlap for about a tenth of a second, or less, depending on the erase current rise time and the tape speed.

Second, when the new recording ceases, either by cutting the record function or by pausing the tape, it returns to the original recording and, recognizable as the familiar annoyance enjoyed by all videocassette users, an un-recorded gap is left due to the timing of tape passage between the heads, except that in an audio recorder it only lasts for a fraction of a second. Still, that's an intolerable flaw in a recording, and the only way to fix it is to continue to assemble-edit the whole program to the end.

A few years ago, a colleague was fortunate enough to operate a Philips Miller recorder; a system which engraved an optical replay track through a black layer on clear tape. Although it could not insert record, it gave his record producer the same edit-capability as photographic film, a substantial advantage over competitors. Later, he used one of the first insert-edit magnetic recorders, which gave his company a similar leap-frog advantage. This machine had an ingenious way around the insert recording problem. The tape was lifted on and off the erase and record heads in a curved path by a mechanical system which engaged the tape on to the erase head last, and then disengaged it first, resulting in a clean entry and exit when inserting a new recording section.

Early magnetic recorders without a facility of this type made a relatively slow assemble edit which gave a double dose of high frequency bias, tending to partially erase the high frequencies and audibly distort the transition.

Music recordings and soundtracks cost thousands of dollars to produce, and their royalty and distribution value is very much higher. When the ultimate value of music CDs or movies is considered, the cost and complexity of production hardware is a relatively small problem. Recorders with insert capability employ precisely timed, smoothly enveloped, sequential erase and record head switching to avoid overlap and gaps when entering or leaving an insert edit. Both functions are switched fast and the transitions are so clean that edits can be made many times on the same portion of the tape without noticeable effect.

But it is the monitor selector panel that enables the operator to sample direct and playback sound, and match levels and quality before making an undetectable electronic edit.

ZERO LEVEL AND LINE LEVEL

Level meter readings all relate to a reference level, and on the VU meter scale that's 0 dB. However, zero VU also indicates a chosen line level, +4 dBm more often than not (1.228 volts in 600 ohms, refer to the decibel table on page 70), so in this case, tone to zero VU is actually tone at +4 dBm in the output line. Domestic line level is normally half a volt in 10,000 ohms. This can lead to some noise and distortion conflicts where domestic and professional equipment is mixed unless level balancing is performed using gain boosters and loss pads.

All line level inputs and outputs are aligned to the same line level so that the condition of unity gain is maintained throughout a system, and so that the monitor selector and all program sources and destinations will see the same level without unexpected changes when switching between sources. Recorders in a studio are all aligned to produce the same magnetization on the tape, which replays to produce line level at all the outputs.

RECORDER LINE-UP TONE

When making an original recording or final mix, it is customary to record a few seconds of tone at zero level on the front of the tape, so that a transfer operator will have a reference as to the program level. This is necessary to instruct a future transfer operator what level is actually on a tape, and it avoids ambiguities like whispers thundering out at full volume if the whole program is at an intentionally low level. But there's a catch. On multi-track tapes, it's a good idea to record the reference tone on all tracks simultaneously, or insignificant tape speed variations will display a phasing effect when tones on all the tracks are heard on the speakers at the same time.

The "line-up" tone recorded on the tape can also be referenced to recorder replay output, because the oxide coating, even on tapes of the same manufacturer, varies from tape to tape in sensitivity and optimum bias setting. This method gives an assurance that the recording will play back at standard level on any other playback deck.

TAPE RECORDER ALIGNMENT

The level and response characteristic of a magnetic recorder is determined by a calibration tape or film, which has several sections which are used in sequence. The replay is aligned first, then the record function, because both are ultimately referred to the playback calibration tape.

Step one is to align the playback head azimuth, if it is adjustable.

Don't proceed with this alignment unless you have the right calibration tape, as listening to the HF response of music tapes is not accurate, and will lose what alignment you have.

The azimuth section is a high frequency tone, 12 or 15 kHz. It is for adjusting the mechanical perpendicularity of the head gap to the direction of tape travel, and it always has to be referred to the actual calibration tape for the record/replay characteristic chosen, as it is critical. Next is a short section of middle frequency tone for establishing the calibration tape's output relative to line level output (it probably won't be the same), and this is followed by the same tone 10 or 14 dB higher, so that playback peak level distortion can be checked. This is necessary, as we will see, because it validates the record peak level distortion measurement at the end of the alignment.

The replay frequency response comes next, and is at reduced level so that any compression effects inherent in magnetic recording will be minimized. Playback output is read as a flat response and adjusted if necessary.

It should be mentioned at this point that there is a variety of record/ replay characteristics in use all over the world, and that a test tape must be used which not only replays the desired characteristic, but is the same version as the standard tape used by the people to whom you intend to send your programs. Making calibration tapes is a specialist operation; they can't be reliably copied because the copy recorder will add its own errors of response and azimuth alignment.

To align the record section, first record an azimuth test so that the head can be mechanically aligned to match your calibrated reply head. The playback alignment is finished and should not now be altered, or it will be necessary to start over. Azimuth can be adjusted while observing simultaneous replay and peaking the adjustment for maximum output. If the replay on your machine comes off the record head, this alignment has already been done.

Adjust the bias by increasing it until playback level just starts to fall on a 1 kHz tone. A slight fall in output gives assurance that oxide variations will not result in an under-biased recording which may introduce distortion. Repeat at 10 kHz to verify that the high frequencies are starting to fall at the same bias setting. Some technicians prefer to set bias for maximum playback output. On a two head recorder it will be necessary to record a family of bias settings, then play them back and choose the optimum. Metering the bias on an oscilloscope will enable this method to be followed with accuracy, provided the oscilloscope is bridged out so that it won't alter the bias level.

Again, record frequency response is done at about -8 dB, as presence of harmonics and magnetic compression will invalidate the measurement. Record a frequency run and check that you have a flat response. Record gain is adjusted so that the whole recorder has unity gain: Input—Record—Replay—Output.

A peak level record distortion test is made at plus 10 dB at 1 kHz. It probably will not be as good as the calibration tape, because your record level may be higher, but this does not mean you are in trouble; peak level distortion up to 5 percent may be no worse than the work of other program producers.

The frequency response of a playback amplifier is subject to a response correction curve in the same manner as a phono pickup, but choice of recording characteristic will depend where you intend to market your tapes. To avoid mistakes, it is advisable to choose one magnetic record characteristic and stick to it throughout your studio. The record level on your tapes is subject to the same recommendation, and while it is based on every recorder you own meeting the peak level distortion specification of 3 percent maximum, with a recorded noise level 65 dB below peak, or better, it is just as important to match the same levels as your associates in the industry, to maintain a standard.

As might be imagined, chaos will result if different recorders produce tapes at different levels within your establishment. In an ideal situation, any of your tapes or magnetic films should be capable of being updated on any other insert recorder, or inter-cut by an editor. It can be done.

Chapter 8.

Mixers

INTRODUCTION TO MIXERS

There is great variety and purpose in mixers; Music recording and mix-downs, Television and Radio, Motion Picture sound-track, and Concert, Auditorium, and Outdoor Event coverage. With the exception of systems that call for preset controls, and some dedicated purpose stereo mixers, the standard format has evolved into groups of modular input and output channels. A basic four in, eight out mixer is shown in Fig. 8.1.

Each input channel has pre-set input level control, often accommodating inputs from microphone level to line level, with appropriate impedance matching. This is followed by linear (sliding) channel faders, low, mid, and high frequency equalizers, and one or two effects faders splitting off to ancillary equipment like the **reverb generator** for sound enhancement and effects, or **foldback** to enable performers to hear the program or parts of it for cues and timing.

The input channels, which usually number from four up to about forty eight are directed to between two and twenty-four output channels, depending on the equipment that follows, and these have their own master faders and insert points for signal processing modules, such as equalizers, compressors, filters, and limiters. This is not to say that insert points and/or processing modules would not exist in every input channel. Sometimes for example, music recordists employ a volume limiter in each source, before the mix, to give to each channel independence of the peak limiter modulation envelope of the others.

The Output Matrix is the selection mix where the input channels are directed to the appropriate output bus. Re-insertion of the effects channels into the chosen output bus also occurs at the output matrix. At each output bus, speaker and level meter monitoring follows, sometimes with an insertion point for optional processing equipment before or after the master fader.

Effects returns do not necessarily have to be included in the same output buss as the input channels from which they originated. This demonstrates the flexibility of a matrix with multi-channel outputs, which can not only produce a stereo or quad set of signals, but also enables producers to make an updatable final mix; leaving them the freedom to re-balance at any time, producing a mix-down with more or less reverb, or a different balance of music, effects, and dialog, or to generate a movie soundtrack mix of music and effects only, for foreign language dubbing.

Although layout and style are the province of individual manufacturers, the group shown on the right of Fig. 8.1 is divided into output masters, with Master Gain control and VU meter, and a monitor mix, which can be set up with tone or pink noise before the session commences, so that not only will each output channel be correctly related to the others, but also the monitor level will be correctly set. For reasons that are made clear at the end of Chapter 6, monitor speaker levels need to be consistent and repeatable to enable accurate balance and equalization judgments to be made. If required, the monitor mix can be used to re-balance a playback of the recording, and on other occasions a separate playback mixer is used.

Figure 8.2 shows circuit details. **Output matrix** switches select the output channel to which each input channel and its source will be directed, and this in turn determines the track on which that source will

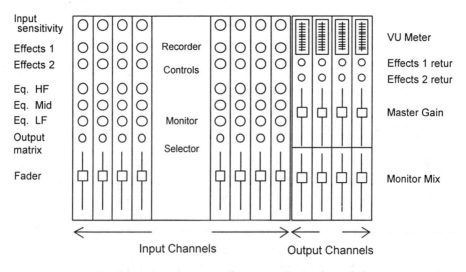

Fig. 8.1 Mixer input and output channel modules.

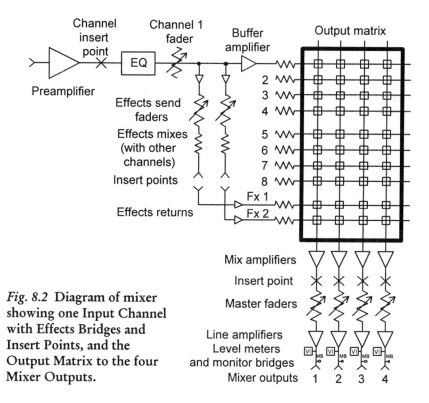

Fig. 8.2 **Diagram of mixer showing one Input Channel with Effects Bridges and Insert Points, and the Output Matrix to the four Mixer Outputs.**

be recorded. The mix resistors before the matrix isolate the input channels from each other to maintain an acceptable load impedance. The master faders are not necessarily all set the same, and can be used, for example, to fade, or balance, an entire section of the mix. Like most faders, the masters are initially set up to "hold" 6 or 10 dB, and that point is marked "0" (zero).

The zero point on a fader is to allow the operator latitude to fade down to infinity, or fade up a few decibels if required. Management of levels is one of the most important parts of mixing, and when all faders are on zero, and the input sensitivity set to line level, an ideal mixer has unity gain with a line level input, at all insert points, effects mix send and returns, and main outputs.

The **buffer amplifier** shown at the end of Input Channel 1, in addition to its loss recovery function, is necessary to stop Effects signals feeding back through the mixing resistors at the output matrix, and causing instability at the effects generators, because even though the effects processors have unity gain, they will certainly ring, at the very least, if the input is partially coupled to the output.

159

Mixer design protocol safeguards the system against noise and distortion by setting minimum and maximum levels from input to output. In Chapter 4, Fig 4.6 shows the relationship between losses in the form of mix busses, bridges, and attenuators being balanced by corresponding gains at preamplifiers, and bridging, buffer, and mix amplifiers, so that although the level varies within the mixer, it is always a recoverable signal. The final buffer amplifier after the mix is also required to provide enough output power to drive the level meter and a 600 ohm line.

Level meters take many forms. Some operators prefer a peak reading meter, but the traditional VU, or Volume Units meter (Fig. 8.3), is a one volt AC meter with standardized ballistics designed to lag, or read lower than, transients (short duration peaks) by 10 dB. The bridging resistor correctly ranges it for line level.

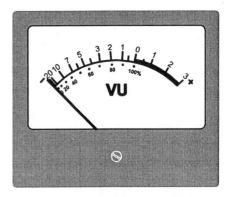

VU meters are identified by a yellow scale, peak reading meters by a black scale with white lettering and pointer. Reading a VU meter is spe-

Fig. 8.3 **VU meter.**

cific. Program that is intended to fill the available dynamic range is peaked to zero on the scale, with the occasional higher program peak exceeding zero VU but not not exceeding +3 dB and running off the scale. Speech has twice the average to peak ratio of many other program components, so drama dialog is often held around minus six VU to leave headroom for high level events. Six dB on this meter is a ratio of two to one, since it measures program level in volts, not power (page 66).

THE DIFFERENCE BETWEEN SPEECH AND DIALOG

Speech from a close microphone contains very little room reverberation. If the person speaking is very close to the microphone, bass content may be higher than is natural, requiring roll-off if the microphone and speaker are both full range. Speaking at a normal distance of twelve to eighteen inches usually produces ideal speech quality, often needing no equalization. Dialog, in the sense that it is part of a drama play, as in theater or movies, is generally picked up under relatively adverse conditions; distant microphone pick-up, a great variety of acoustic conditions, some with high degrees of echo and reverberation of undesirable frequency spectrum, and noises-off that may have nothing to do with the action.

Like the slowest ship setting the speed of a convoy, medium dialog quality has to be catered for: in many cases, screen dialog that was recorded in different places will be inter-cut, and has to be matched. Dialog that can not be improved is either edit-replaced with **wild track**; dialog recorded on the set without the camera rolling, so that a better microphone aspect can be used, or **post sync.** recording; dialog recorded in the studio to a screening of the scene, with the original sound as a **guide track**. Whatever the origin, dramatic dialog, usually for Motion Picture Soundtrack, needs further processing to ensure the listeners perceive it as reality (Ref: Preparing for Transfer Errors, Chapter 5).

One of the processes which assists matching of different sound qualities is **bandpassing**; limiting the effective frequency response of the program. It is desirable for other reasons too, but has to be done within limits depending on the results of other processes (Fig. 8.10). High sound pressure levels in big audioriums are unnatural and hard to listen to if dialog contains heavy bass content and third order harmonics in the high frequency band. The effective range of movie dialog is within the frequency band 70 to 6,000 Hz, a balanced spectrum according to the 630 Hz formula given in Chapter 4, and discussed in relation to equalizer/filter combinations on pages 169–172.

DIALOG PROCESSING

Dialog is very different material to handle in a mix compared to music and sound effects. Speech recorded under less than ideal studio conditions requires the opposite environment to music, which calls for bright, reflective acoustics to reinforce the high frequencies and provide sustain. Drama dialog is recorded where possible in non-reverberant conditions. This is to control echo, and make it possible to inter-cut different shots recorded in a variety of conditions.

When the dry-recorded and edited track is balanced in the mix, dialog processing brings it back to life. The opposite is necessary with close-microphone work in a studio, where little or no processing is used. If dialog intended for editing is recorded in a normally reflective situation, intelligibility will be masked by echoes and reverberation. This does not occur to the same extent with direct binaural hearing, because that is backed up by cerebral processing.

One of the main problem areas is excessive bass, because room reverberation predominates in the lower frequency bands, mid and high frequency reverberation being lower, and easier to control. The majority of indoor recording venues do not have a flat frequency absorption characteristic, but that is an aim of sound stage design, and some methods of achieving this are detailed in Chapter 9. For similar reasons, sound destined for auditorium or theater reproduction needs additional pre-processing if it is to survive the reproduction environment, which adds the product of multiple reflection paths to the direct sound.

When the soundtrack is assembled, scene to scene matching is achieved by equalization and level balancing. In order to make possible inter-cuts of dialog recordings of different perspective, or components of original sound on camera and replacement dialog recorded in the studio, two or more tracks are laid, so that each type of voice recording can be separately equalized and balanced with its individual channel equalizer before combining at the dialog group sub-mix. The dialog is then sent to the dialog chain for processing before combining in the final mix via the dialog group return fader (Fig. 8.4).

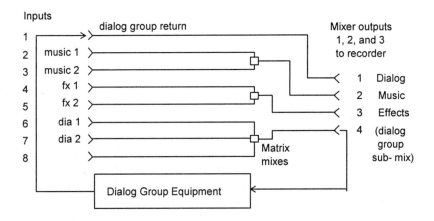

Fig. 8.4 The dialog group sub-mix uses a spare output channel, and returns via an input channel fader for inclusion in a single or three track mix.

The dialog treatment process, which could not be done manually, brings back to life dialog that was recorded dry, without natural reverberation, and lifts it out of distance and ambient noise to create the familiar smooth voice quality that characterizes a good movie sound-track.

DIALOG EQUALIZER / COMPRESSOR COMBINATION

The first item in the processing chain is the main dialog equalizer, which rolls off the bass from about 600 Hz down, to the extent of between 3 and 6 dB loss at 100 Hz, and then applies sufficient mid frequency boost between 2 and 3.5 kHz to bring the dialog to a central balance (Fig. 8.5). It would be possible to use the individual channel equalizers to apply the dialog equalization characteristic, but it is preferable to do it later with a single dedicated equalizer, as it keeps the channel equalizers referenced to a central, flat response position, making it very much easier to achieve a balance between the tracks prior to combining them (Fig. 8.6).

Traditionally, dialog is then compressed two to one; the top 20 dB of the program into 10 dB, but usually not a higher ratio, as this would compromise the dynamic range required for dramatic program. Also, soundtracks are sometimes further compressed in later generations like copying and broadcasting.

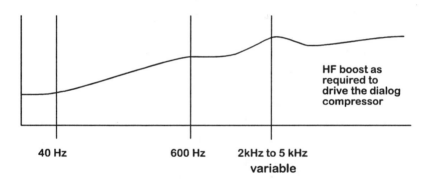

Fig. 8.5 Dialog equalization curve.

Dialog is compressed for several reasons. It contains high level transients that would be difficult to listen to at theater sound pressure levels, and it also has essential low level components that cannot afford to be buried under ambient auditorium noise, or the rubble of background sounds that are also part of the sound-track, added at the final mix. Dialog is the message part of theatrical product, and if the audience can't hear every word, then a great part of its dramatic purpose is lost. Producers go to great lengths to coach their cast in good dialog presentation. It therefore warrants all the technical back-up it needs to maintain the value of the product.

The volume compressor reduces the dynamic range of a program. It is

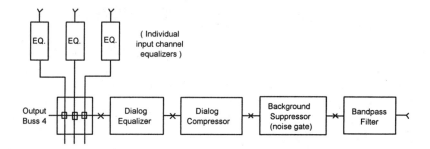

Fig. 8.6 **Elements of the Dialog Processing Group.**

an electronic attenuator controlled by the DC envelope of a side-amplifier fed by the compressor output. The compressor senses the effect it is having on program levels, because it is driven by signal it has already processed. Compressors are backward-acting. They monitor the output to correct the input, but are different to limiters in that they reduce dynamic range without setting an irreversible ceiling level.

The dialog equalization characteristic of Fig. 8.5 is a volume compressor driver. Its purpose is to control the attenuation that the following dialog compressor exercises on different parts of the spectrum. Much of the effect of the dialog equalizer is actually removed by the compressor, but in doing so, it processes the dialog in a special way.

The combination of equalizer followed by compressor has been called the **Electronic Mixer,** because it automatically smooths and balances the sound, not just reducing dynamic range, but boosting mid-high frequencies when they are lacking in the speech, and holding them when they are too prominent. A similar effect is achieved in the bass, the compressor selectively putting back, to some extent, bass that has been removed by the equalizer, but doing it at the right time, and at appropriate parts of the spectrum.

Voices of different national origin can require different equalization, especially in the choice of the boosted "presence" frequencies, but the principles of dialog equalization followed by compression hold good regardless of voice type.

A straight volume compressor is unsuitable for dialog, as this type of program contains **sibilant** or "s" sounds that possess insufficient power to adequately command the electronic attenuator, yet are capable of going into distortion by peaking beyond the capability of recording media if they are not compressed in proportion to the rest of the spectrum. Even if they don't distort, sibilants out of proportion make the sound "chirpy," and hard to listen to, an effect that can not be reversed by further equalizing. Like many natural sounds, sibilants occupy a wide band of the spectrum, which is why they can't be equalized or dip-filtered out without spoiling the quality of the program. This introduces the first rule of signal processing:

NEVER EQUALIZE OR FILTER TO REMOVE NOISE OR PROBLEM
SOUNDS AT THE COST OF SPOILING THE SOUND QUALITY.

Fortunately, there is a direct and non-destructive way to to solve the problem (Fig. 8.7). Dialog compressors use an equalized side-chain to put the power spectrum of sibilant sounds back in control. The program doesn't actually pass through the side chain, so it is not spoiled by boosting the side chain in the frequency band where sibilants occur. De-essing, as the process is called, has a second and particularly valuable effect. It greatly improves the quality of dialog processed by the equalizer driven compressor, producing a smooth, pleasing dialog sound free of uncontrolled "esses," that sounds natural at high levels in large auditoriums, and at low levels in the presence of ambient noise on radio and television.

The dialog compressor has further dynamic equalization. Its attenuation curve is not linear at all frequencies. When signal level is below the threshold at which compression commences, the compressor has a flat response. But as input level and therefore attenuation increases, the center of the spectrum is attenuated later than the two ends. This augments the middle frequencies, between 2 and 3 kHz, at higher levels. This is the

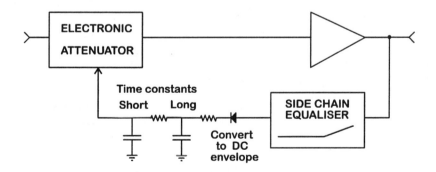

Fig. 8.7 Dialog compressor block diagram, showing the side-chain's equalizer and dual time-constant envelope.

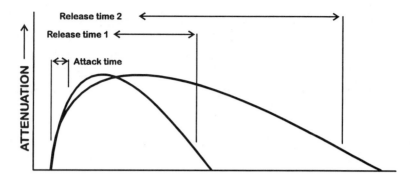

Fig. 8.8 Release time 1 recovers fast following transient peaks, while release time 2 returns average program level changes more slowly.

region where dialog intelligibility resides, and as levels increase, the "presence" components of speech are reinforced, preserving perspective while maintaining acceptable dynamic range.

Another feature is dual attenuator release time constants (Fig. 8.7) which help to prevent the rapid changes in the attenuation envelope audibly pumping the level up and down, and it does it so effectively that few listeners are aware of the compressor's operation. The effect is shown in Fig. 8.8.

It will not be obvious until it is experienced, but using the dialog equalizer after the compressor, so that its effect is unmodified by compression, will fail to smooth the dialog, and may instead produce a harsh voice with peaks that are fatiguing to listen to. In this arrangement it will not be possible to use as much presence equalization as is needed to bring out the essential components of the dialog sound, and there will be no Electronic Mixer effect.

The order in which non-linear processing devices are connected is critical, and experimentation will serve to illustrate that an unsuitable signal processing format will produce undesirable effects, calling attention to the second rule of signal processing:

DON'T OVER—PROCESS SOUND. IT MAY BE GIVEN FURTHER PROCESSING LATER IN BROADCASTING OR COPYING.

WHERE NOT TO USE DIALOG PROCESSING

The full extent of the dialog treatment just described is totally unsuitable for music or general sound effects. If it is left until after the final mix to use equalization followed by compression, very limited treatment can be given, because enough to satisfy the dialog will seriously damage the music and sound effects. So dialog is sub-mixed in its own group, processed, and fed back into the final mix via the dialog group return fader, as was shown in Fig. 8.4.

BACKGROUND SUPPRESSION

The last two parts of the dialog chain deal with trash removal. In order to match inter-cuts of different shots from the same scene, a background suppressor, or noise gate, can be used to reduce the dialog background level.

The background is seldom stripped completely, even though some continuous "atmosphere," or ambient sound that is normally recorded after each dialog scene is shot, can be added back to make the mix realistic. Dialog sounds unnatural if its background is reduced too far, as it still contains the original background noise at full level behind the words, and only the spaces between them are attenuated.

Even if the background is a noise that should not be there, consider the combined rules of signal processing, from the point of view of the theater audience hearing the finished product:

IF IT DOESN'T SOUND GOOD, DON'T DO IT.

In practice, the background can be reduced from 6 to 9 dB without ill effect. Like all processing, it is better if it is undetectable to the casual listener. Dialog that has the natural over hang of each syllable clipped off can end up sounding "wooden," like a vintage movie. But if background suppression is performed with reserve and skill, then most audiences will be unaware of the process.

Background suppression is carried out after equalization, when low frequency elements of the sound have been attenuated, and presence frequencies boosted; after all, it is best done on the spectral balance that will be finally heard. However, it also comes **after** the compressor, because the two thresholds of compressor and noise gate can be set up so that they do not overlap, as shown in Fig. 8.9. Also, because the equalizer has a direct effect on the compressor's action, the two should not be separated by another processor. All non-linear processors "envelope" the sound in a particular way, and the skill of using them is to have them do

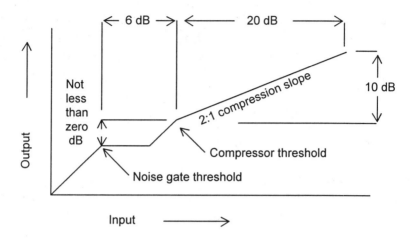

Fig. 8.9 Input/output slope of the compressor/background
suppressor combination.

their job without it being obvious that the sound has been processed. A description of early background suppression methods on page 176 shows how the situation has improved.

BANDPASS FILTERS

At the end of the dialog chain, following equalization, compression, and background suppression, comes bandpassing. The end frequency filter is usually the last item in the dialog chain so that any audible low frequency "envelope" effects generated by the compressor and noise gate will be attenuated, but it would be a valid consideration to place it first, so that the "electronic mixer" acts on frequencies in the balance that will be heard. This is really a debatable point because today's dialog is not bandpassed as severely as it was in the first few years of movie sound, and the processors now work better, but there is another compelling reason for bandpassing last.

The filter is set to rapidly attenuate frequencies below 60 or 70 Hz that could become troublesome in extremes of auditorium response and reverberation error. At the high frequency end, the relationship of harmonics to speech fundamental frequencies was introduced in Chapter 5. In dialog filtering, one of the aims is to preserve the natural balance of voice harmonics and subsonics despite the imbalances caused by equalization and compression. The slope and cutoff frequencies of the bandpass filter are chosen to optimize this balance.

The band limitations imposed on dialog are not used on the overall mix unless it is for a very good reason, as stereo film soundtracks and theater reproducers are capable of handling the extended response of music and effects. These components of the mix do not exhibit the transient peaks and runaway high frequencies that characterize speech. It has long been the practice to compress and bandpass the whole program when it is transferred to 16 mm film, in order to accommodate the anticipated wide variations of program type and the limited dynamic range. But on the wider latitude medium of 35 mm film, material that won't get into trouble in further recording and reproduction generations is left unchanged after the final mix for auditorium speakers to apply their own bandpass (Chapter 6).

The second harmonic and its adjacent overtones are the character components of stringed instruments and the human voice; higher harmonics to a lesser extent. Dialog high frequency response is tailored by the filter to maintain these valuable frequencies above the presence region that give the voice character, but to control the higher harmonics. Twice 3 kHz is 6kHz, and we want to hear it, but the second harmonics of 5 and 6 kHz can be trouble if they get out of proportion. In the same way that noise exists, but is not a problem unless it is above a certain threshold, so the upper harmonics of dialog have a rightful place if they are in balance. Unfortunately, relatively high reproduction levels, combined with the fact that volume increases are more noticeable at the ends of the spectrum than in the middle, changes the perceived proportion of these harmonics, hence the bandpassing. The subject of just what a filter does is about to come up, and it will be seen that the upper harmonics are not entirely lost; but they are controlled.

Speech components as high as 10 kHz and 12 kHz don't contribute to intelligibility in a large auditorium; quite the reverse, as multiple length reflection paths produce cancellations and duplications that would turn high frequency transients into mush. This is not true of music, as it doesn't have the rapid-fire data stream of dialog, and a band or an orchestra is naturally loud, radiating sound in all directions and making full use of complex reverberation. Thus dialog radically differs from music, and calls for quite different processing.

The 630 Hz tool, introduced in Chapter 4 as the logarithmic center of a balanced spectrum whose end frequencies product is 400,000, also has a bearing on the selection of cut-off frequencies when bandpassing other program components. Not that mixers whip out the calculator every time they make a filter decision, but it's useful to consider the 630 Hz guideline when using filters to create perspective for effects or dialog, and sound effects simulations of radios and telephones; anything that would have a restricted response if it was part of a movie scene. It makes them sound real.

The 630 Hz formula is also a very good guide for setting the dialog group bandpass filter. For example, if the high pass is set between 60 and 70 Hz, then the effective upper cut-off frequency might be 400,000 divided by, say, 67, which equals approximately 5,900 Hz. The voice sound will be balanced. It's not that straightforward in practice, because considerable modification has been made to the dialog spectrum, and the **effective** cut-off is going to be higher than the filter setting indicates.

It has long been considered that 6 kHz or thereabouts is the the minimum acceptable high frequency limit for theatrical dialog, so it's a good figure to aim at, because we have just seen that it may not be a good idea to exceed it by very much. A filter cut-off setting between 4 and 6 kHz is often used. The term "cut-off" defines the 3 dB loss point of a filter, but in conjunction with other processing events like the equalization plus compression combination and the slope of the filter, mid-high frequency boost can make the **effective cut-off frequency** a lot higher.

When a mixer experiences difficulty with some voice types, it is not uncommon to resort to using more than usual mid and high frequency boost to drive the compressor harder, and then it is possible to set the filter as low as 3 kHz. The end result may still have an audible content up to 6 kHz, which is one octave higher, if 12 dB of HF boost was used prior to compressing and filtering with a slope of 12 dB per octave. That's an extreme case presented to illustrate the point, and there's nothing quite like hearing it in practice, but Fig. 8.10 attempts to show how HF boosted program and a filter slope of 12 dB per octave will **sum** to produce a 3 dB loss point which is appreciably higher than the filter setting. The 70 to 4,500 Hz bandpass filter will probably produce cut-off frequencies of 70 to 5,700 Hz, because of the HF boost produced by the dialog equalizer/compressor combination. Furthermore, it is not the whole dialog that is thus extended, but the lower levels, because of the action of the compressor in controlling just where and when the equalizer has an effect.

Concept 1: Filters don't cut off sharply; they have a defined slope.

Concept 2: The actual cut-off frequency is determined by the filter setting, its slope, and the equalization used.

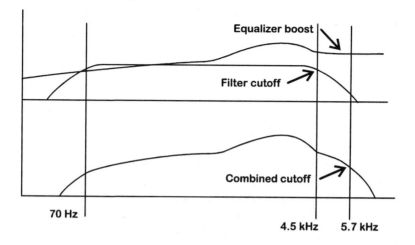

Fig. 8.10 The dialog response curve at the top of the chart is superimposed on the response of the filter, showing the 3 dB points. The bottom curve represents the sum of the two responses, demonstrating how the HF boost has extended the effective range.

The dialog processing chain is a set-and-forget tool, otherwise a sound-track would keep changing character throughout its length. It's always a good practice to log the settings so that they can be re-established if they are changed. This valuable practice also enables others to update or continue a job where necessary.

In educated use of the dialog chain, its not the position of the controls, but what it sounds like that counts, because the compressor selectively removes a great deal of the equalizer's effect, and the equalizer modifies the filter's effect.

Concept 3: Where a filter follows an equalizer/compressor, its not the position of the controls that counts, but how it sounds.

The process in which the equalizer and compressor interact augments upper harmonics. They would fly away out of proportion were it not for bandpass filtering. As described in Chapter 5, it's the harmonics that give a voice it's character, but if they are not in natural balance with the fundamental frequencies, they will sound like distortion because they are made of the same stuff.

Concept 4: Some valuable processes would be unusable without filtering to correct the harmonic imbalance they cause.

DIP FILTERS

A useful adjunct to the dialog chain, in fact one that is built in to many active bandpass filter units, is a pair of dip filters. This type of filter sharply attenuates an narrow slice of the spectrum in much the same way as the noise and distortion meter of Fig. 5.5. The width of the slice removed from the spectrum is variable, but it can be set fine enough to be undetectable in ordinary program.

Two such filters are often needed to remove both the problem frequency and its nearest harmonic, or two problem frequencies closely spaced. They can often save a dialog recording which has been marred by bird or cricket noises. They won't fix sibilants because these occur over a surprisingly wide band. That has to be done dynamically with a purpose designed compressor.

In a well equipped mixing suite, a dialog "cleaning" bay includes a tape loop recorder on which to copy a section of a troublesome dialog passage so that it can be analyzed off-line with the dip filter and background suppressor. The settings established are used when the mix is continued on-line.

Since soundtrack production invariably uses insert recorders which can run back and pick up a recording with all component tracks in sync, it's a matter of pausing the mix, setting up the the filters, and continuing with the new settings. Alternatively, the mix can continue unbroken, and trouble spots returned to for updating without disturbing music and effects mixes on the other tracks.

POINTS TO CONSIDER WHEN MIXING

There are two areas where the physiology of hearing can cause trouble if the novice is unaware of it. One is the tendency for levels to creep upward as the mix progresses, due to listening fatigue and the mesmeric effect of concentration on the job. The other is similar, but stems from different causes. It is the tendency to increase mid and high frequency boost as the mix progresses. In the first case, the novice mixer may find that levels are gradually increasing as the ears becomes used to the volume, and in the second, dialog can be made brighter and more strident as tension increases and the perceived presence balance changes. This is due partly to listening fatigue but also because perception of presence may change under the duress of a difficult mix that requires long periods of concentration.

Routine observation of the level meter will help in the first case, although the meter is a guide and a starting point and it doesn't replace the ears as arbiter of comparative levels and sound perspective. An established method of reading the VU meter, use of a standard monitor setting, and a log book to help consistency of equalization, make up some of the ways an operator uses initially to achieve a uniform result. To arrive at a formula for a particular mix may take some time, but most people expect to run more rehearsals during the early stages of a mix, giving them time to establish a feel for the program.

When starting a new production, it is advisable to take it slowly for the first couple of hours, then take a break and review the first part; comparing what was done with the current section being mixed. Another aid is what some seasoned mixers habitually do, and that is to log all their settings in a notebook, maintaining mixing uniformity by periodic reference to them.

Repeatability of level and equalization settings is also important because the mix may have to be up-dated at some future time, perhaps by someone else, and there's nothing as useful as a log book which stays in the mixing suite for the purpose.

Rock and roll mixing, as it has been called, is a great asset, because not only inserts, but complete mixes can be done section by section, edit-assembled until finished. However, it is logical that a mix will be more fluent, and have a higher theatrical standard, if one or two good rehearsal runs are made first, and then a "take," without stopping, or at least a minimum of pick-ups, to record the final mix.

THE EFFECTS CHANNELS

Reverb, echo, and the various audio delay and phasing generators available for different effects can all be included in the mix via the **effects send** and **effects return** controls. Turning back to the first three pages of this chapter, it will be seen that the layout and block diagram of a basic mixer show the effects channels, but without a lot of detail, in the interests of clarity. Very simply, many effects bridges can be switched **pre-fade** or **post-fade**, in order to provide maximum flexibility.

The reverb generator is perhaps the most used effect, for it will work in so many different perspectives; full reverb of various decay times, subtle effects, and as a matching tool for interior-exterior or dialog replacement inter-cuts. Pre-fade reverb can be a potent off-stage effect, or post-fade reverb can simulate many situations if combined with equalization. Compression is sometimes used in a reverb generator drive chain to apply a ceiling to the process for the sake of the reverb generator's optimum performance and also if it sounds better that way.

The simplified mixer block diagram shown in Fig 8.11 shows how an open-reel or tape-loop recorder with simultaneous playback capability serves as an effects generator. It can simulate reverberation in the absence of more specific equipment, producing limited but acceptable results on most sounds except percussive instruments or effects, where the repetitive nature of the simulation becomes apparent.

Merely connected to the insert point, the recorder won't produce anything except a single echo related to the tape speed and the distance between the heads. To make it reverberate, that is, generate multiple echoes which feed each other and build up a continuous effect, the output is fed back to the input via a potentiometer which acts as a reverb time

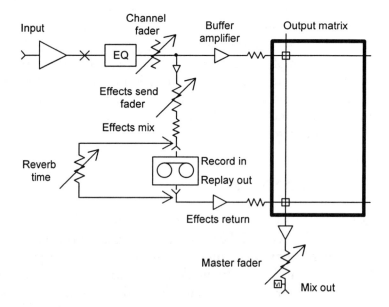

Fig. 8.11 Three head tape recorder as effects generator.

control. Reverberation time is by definition the time it takes for a sound to decay by 75 percent of its original level.

The by-pass is necessary to establish positive feedback from tape delayed output, back to the tape recorder input. This is because there is a buffer amplifier before the matrix mix which prevents the effects channel feeding back to the input channel to cause trouble.

Commercial reverb generators include electro-mechanical and electronic delay-line types, which have flexible and quickly accessed variable settings, often combined with other effects. Even the older spring type reverb generators sound good if they are not driven too hard, but because of the low mechanical levels at which they operate, they need to be installed in isolation from vibration and magnetic fields. Tape loop delay is sometimes used to drive reverb generators of all types to augment their effect.

In the absence of any of any compact purpose-built reverb generators, the opportunity to set up a live room as an echo chamber should not be overlooked. The chamber consists of a suitably live area, like a concrete stair well or a bare passageway. An amplifier and speaker in the area is driven by the effects **reverb send**. Then a microphone separated by baffles from the speaker drives the **reverb return** via a microphone preamplifier. The biggest problem a live echo chamber encounters is ambient noise entering the area from outside, but the results can be as good as or better than any of the compact systems.

DIGITAL SOUND PROCESSING AND EQUIPMENT

The advances in all areas of sound since the introduction of digital techniques in no way diminishes the style of the systems and techniques described in this chapter, because the intent is the same, and without the research and evolution which is embodied in them, sound today would probably have taken different directions.

Digital and computer processing can no doubt simulate all the features of the dialog chain, but the principles have been described in the original analog terms for simplicity. The main point is the value and necessity of these processes, not how they are achieved.

There are two factors which digital sound technology offers beyond the capability of analog systems. First, some things are possible that just could not have been done at all with analog equipment. An example is digital sound encoded in a small area of analog/digital compatible Motion

Picture prints, and not just sound, but five discrete channels of a quality that surpasses the best that analog can achieve.

The other is that most things are done far better using digital technology. Many people in the sound industry may be unaware that a digitally processed volume compressor was developed several decades ago, for which very good results were claimed. Perhaps it did not come into general use because of the difficulty of manufacturing such systems with the technology of the time, but it does demonstrate that few things are really new; they're just waiting for better materials and know-how.

UNLOCKING THE NOISE GATE

In 1960, I was employed in the Film Production arm of a broadcast company with a highly regarded engineering department. The engineers were asked to solve a background noise problem which was spoiling dialog recording at an outdoor location, a town set built for the production in hand, complete with a flowing river-crossing, streets, and buildings.

Within a very short time, three engineers brought in a prototype of the Anticipator, so called because it was designed to "wake-up" just before the first wavefront of the dialog. It consisted of a tape recorder to provide early warning of the event by delaying the program sound via off-tape replay. The attenuator was a network around a relay triggered by direct program before the delay. The attenuator was suitably buffered so that it gated the program level smoothly up and down.

On the trial run the Anticipator's performance was as perfect as its design and construction, but there was only one catch. The background noise it was asked to remove consisted of the roar of the pumps which made the river-crossing work. The noise was about the same level as the dialog, and to my regret, the project was discontinued.

The abandoning of the device did not mean it wasn't a great idea superbly executed, and history will also testify to the ingenuity of the Dynamic Noise Suppressor developed of the 1930s, which limited the bandwidth of the program below a pre-set threshold.

Later, the first background suppressor I used consisted of a neon lamp driven by a side amplifier, which illuminated an LDR (light dependent resistor), which was the top half of a fixed attenuator pad. It cost almost nothing to make and was easy to set up, with the threshold and degree of attenuation variable within wide limits on two poten- tiometers. It was used for several years until the price and size of commercial units came down.

Neon tubes ignite and go out very fast, and would be intolerably noisy were it not for the attack and decay characteristics of the LDR, which goes high resistance in the dark. When illuminated, resistance falls fast and smoothly, recovering more slowly when the light goes out, in an ideal audio attack and decay envelope characteristic.

LDR's have a light/dark resistance range of about fifty ohms to one hundred thousand ohms, depending on the type, and can be parallel connected to suit circuit impedance. However, they should not be asked to pass much current in this application, to preserve linearity and low distortion.

These days, it would not pay to make your own noise gates, because compact multi-channel units are available at the right price, small enough to place a whole row of threshold and attenuation controls at your fingertips.

Chapter 9.

Acoustics

ROOMS AND ACOUSTICS

The acoustics and noise isolation techniques described in this chapter are only the most basic examples of the many ways to approach the subject. But they will show that improvements to the acoustic environment are well within reach, and that there are many ways to achieve an economical result.

The need for sound absorption is not universal. Some reverberation is essential, as in music presentation and audience empathy in auditoriums. To avoid spoiling the essential character of large public venues, discussion is directed to relatively small rooms such as in homes, theaterettes, and sound studios where the two techniques of **Sound Isolation** and **Acoustic Control** are appropriate.

REVERBERATION AND THE ABSORPTION SPECTRUM

The acoustic environment has a profound effect on our well-being, particularly in the workplace and at home, where we spend a great part of our time, but the same comment applies to studios and auditoriums.

A factor we can call the **acoustic climate** is determined by the reverberant properties of the room, or the sum of its sound reflections. An echo is a single reflection; reverberation is the complex made up of multiple and repetitive reflections within a room.

Reflectivity of a material or structure is the inverse of absorption, the factor standardized in reference to an open window, the ideal non-reflector. Components of the room, its size and shape, combine to make its acoustic climate.

SOUND ISOLATION AND ACOUSTICS

The two groups of influence in rooms are separate and distinct, as is their cause and treatment (Fig. 9.1). **Isolation** deals with prevention of sound transmission by air, through walls, floor, and ceiling. Another area of sound passage, especially in steel or concrete buildings, is by conduction through the building frame, and the major source of noise in this mode is from impacts directly onto hard surfaces connected to structural members. Door and window frames present a loss of isolation integrity, and in essentially soundproof places like studios, care is taken to contain air leaks, for example, by sealing door and window frames to the walls on both sides.

Acoustics is concerned with the way the room's shape and materials modify sounds inside the room. Natural sound is pleasing to the senses because it's what we're used to.

If absorption is one of the main factors in exercising control of a room's acoustics, then it should be in proportion across the audio spectrum. The **absorption spectrum** is less evident in open space and has much

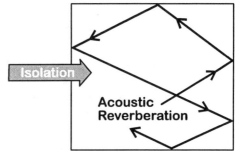

Fig. 9.1 Acoustics and sound isolation are different effects with different causes.

the same effect at all frequencies. Reverberation, as distinct from echo, is the result of multiple relatively short-path reflections within a room which continue until they run out of power, due to the small loss acquired at each reflection. Hard parallel walls reinforce all frequencies, and while soft furnishings absorb much of the middle and high frequency spectrum, the average room leaves low frequency reverberation dominant.

An oppressive feeling comes from an environment that is boomy and lacking a balance of high frequency presence. Sound entering a room from outside also has the middle and high frequencies attenuated by its passage through the walls, causing the bass to predominate. Whether an unnatural response spectrum is produced within a room or comes in from outside, the disturbing result is the same.

What is the value of a room with equal sound absorption at all frequencies? It feels open and natural. Like the outdoors, people feel at ease in such a room. Sound modified by reverberant conditions within the room can be remedied by a variety of measures which often differ from normal architectural and decorating practices. None the less, they can be applied to most small rooms without looking out of place or breaking the budget, remaining practical, and above all, finished so that they can be re-painted or re-covered without reducing their effectiveness. While the acoustics can be altered by adding materials, isolation has to be dealt with from the ground up, being totally involved with the structure.

Various practices and philosophies are described in this chapter, but I stress that they are not the only ones of value. The basic tenets of acoustics and noise control remain the same; to elevate quality of living, to make the best use of building materials and designs, and to acquire affordable studio and recording rooms.

EQUALIZING THE ABSORPTION FREQUENCY RESPONSE

The **Inverse Square Law** introduced in Chapter 1 makes it impractical to use the same techniques, simply re-scaled, to govern all parts of the audio spectrum. At least two different methods have to be used to control reflections, overlapping their effective frequency bands in the same way as high and low frequencies in speaker crossover networks.

For many acoustic absorbers to be effective, they must have some dimensional relationship with the wavelengths of the frequencies involved. In accordance with the formula on page 57, the wavelength of sound in air of 10 kHz equals about one and a quarter inches. At 100 Hz, it's more like ten feet.

Soft absorbents like carpet and padded furnishings relate dimensionally to the mid and high frequencies whose wavelengths are between one and twelve inches, but control of low frequencies with baffles around ten feet wide protruding into the room is impractical. So a different technique is used for low frequencies in the form of the **Panel Resonator Absorber**, which works by mass-compliance rather than by dimension.

ACOUSTIC CONTROL BY DIFFUSERS AND RESONATORS

While soft furnishings are effective at upper-middle and high frequencies, there is a method which operates just as efficiently over a wider band, but retains a natural brightness. It is diffusion or scattering, using the principle of **Anechoic Chambers**, where deep fabric-covered fiberglass wedges fixed on the walls and ceiling cancel reflections by appropriately changing their angle.

The system for less specialised rooms is simpler. The vacuum-formed wedges shown in the photograph of Fig. 9.4 are 5" square by 3" deep. In combination with the 3/16" hardboard low frequency panels, the absorption is reasonably constant over most of the audible spectrum. There is an "open air" ambience in the two preview theaters shown in the photograph, and also in the plan view of Fig. 9.8. They were treated with four foot wide mid and high-frequency diffusion panels alternating with low frequency panel resonator absorbers of the same size (Fig. 9.3). The double skin suspended ceiling has a convex shape, being lower in the center, to further disperse reflections.

The rooms are small, but broad spectrum reflection is reduced and standing waves are under control.

While the loose dimensions of the high frequency panels were chosen for low profile appearance and ease of forming, the low frequency resonator absorbers were calculated according to the formula:

$$f = \frac{170}{\sqrt{m.d}}$$

where f is the center absorption frequency, m is the mass of the panel in pounds per square foot, and d is the depth in inches of the air-gap behind the panel.

The same formula for panel center frequency is expressed in the metric equation:

$$f = \frac{60}{\sqrt{m.d}}$$

where m is the panel mass in kilograms persquare meter, and d is the air-gap depth in meters.

If the cavity of the LF resonator is damped with a layer of felt or other soft absorbent material, it becomes a low Q device, and the response spreads over a broader band (Fig. 9.2). The panel absorber can be made to blend with the decor, as its front is flat and can be covered with fabric, which also helps to control high frequency reflections from its surface. The effect of as little as 40 percent of wall area treated is dramatic; far more effective than would be expected from so simple a treatment. Other forms of absorber could be made to work in the low frequency band, but like the Helmholtz Resonator, a box with a tuned hole, they tend to work in narrow bands.

Some remedial systems or procedures are so simple that nobody uses them. An environment improvement concept whose costs are the same as traditional practices are difficult to market, except as a service. However,

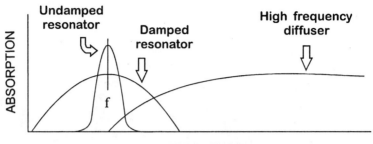

Fig. 9.2 Absorption response of a panel resonator plus high frequency diffusion panel.

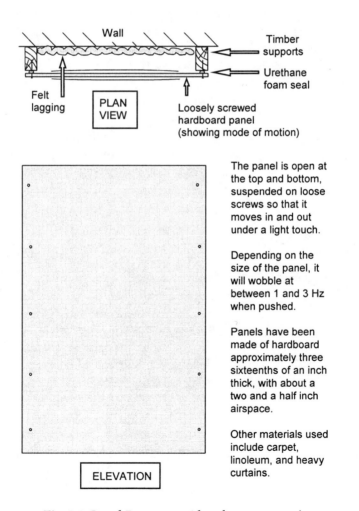

The panel is open at the top and bottom, suspended on loose screws so that it moves in and out under a light touch.

Depending on the size of the panel, it will wobble at between 1 and 3 Hz when pushed.

Panels have been made of hardboard approximately three sixteenths of an inch thick, with about a two and a half inch airspace.

Other materials used include carpet, linoleum, and heavy curtains.

Fig. 9.3 Panel Resonator Absorber construction.

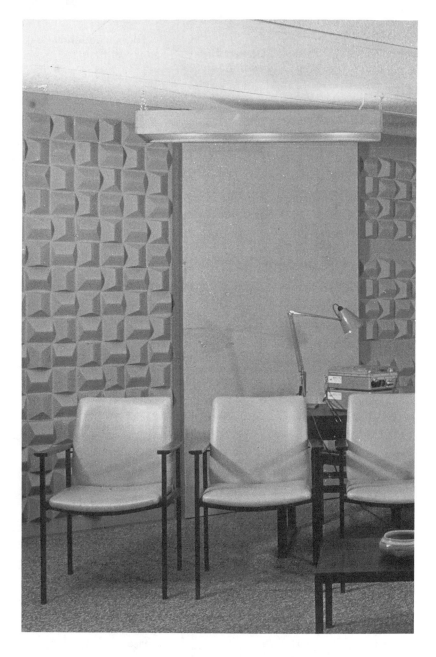

Fig. 9.4 High frequency scattering panels made from 22" square vacuum-formed styrene tiles alternate with fabric covered low frequency resonator absorber panels.

while most acoustic ceiling and wall tiles are designed to control nuisance frequencies, there are a few commercial panels made which cover most of the spectrum. Methods include slotted plywood boxes or panels, and frames covered in heavy, felt-backed perforated vinyl. Shape-cut urethane foam panels, popular in studios, may need assistance in the bass spectrum.

The HF and mid frequency diffusion panels shown in Fig. 9.4 need not spoil the decor of a room if they look out of place. They can be recessed or framed, and covered with an open-weave fabric to conceal them completely without reducing their effectiveness.

Equal areas of high and low frequency panels are a starting point, but the proportion can be varied to suit the need. Like air conditioning, some is better than none, and it would be reasonable to expect substantial improvement if only 50 percent of wall area was covered in the two types of panel at the height relative to seated and standing people, from two to eight feet above the floor.

THE ABSORPTION SPECTRUM SHOULD AIM AT A FLAT FREQUENCY RESPONSE

Understanding of the principle involved will be useful when hanging heavy drapes spaced a few inches off the wall, or having an acoustic consultant fix a low frequency resonance in your studio. Chances are, the specification for remedying the problem will include a fabric covered panel suspended three inches off the wall that is propagating the troublesome reflections.

There is not always a choice, for practical or decor reasons, but it is the main direction of sound travel that determines which walls should be treated. Small rooms in which standing waves cause the formation of nodes of high bass response will be candidates for panel absorbers on adjacent walls, to reduce the low frequency reflection that causes the problem.

Difficult locations like gymnasiums or bare halls need a lot more area treated, and perhaps in these and utility areas the emphasis would be on more controlling the noise spectrum than an ideal listener's environment. Suffice to say that there are different acoustic requirements for different programs. A dance hall or music studio benefits from the right kind of reverberation; in fact it's necessary. But to give a speech or make a recording in such a venue may mean you won't be heard intelligibly.

Should you want to put customers at their ease in a restaurant, or solve communication problems with customers at a fast food counter, there are many acoustic products available for tiling ceilings and walls, using helmholtz resonator and other principles. Most of them are for absorbing the nuisance noise spectrum above 1 kHz, and while some of these commercially available products may be incorporated into any serious sound venue, some form of low frequency absorption would also be needed.

Some materials do not perform as expected. Round fiber monofilament fiberglass woven mat is acoustically transparent, as if it was not there. Wool is effective and is to an extent fire resistant. Existing wall panels will influence the acoustic climate of an area if they are able to flex, driven by sound waves. Speaker users sometimes find their bass response compromised by loosely supported lightweight wall or ceiling panels.

SOUND ISOLATION CONSTRUCTION

Since sound is conducted by hard materials, pouring tons of concrete is no guarantee of containing it. The effective method of isolating vibration of any kind lies in employing alternate zones of dense and compliant character. Consider vehicle suspension. The chassis and cabin present a mass which tends to stay in place through inertia, because the springs provide the compliance. Therefore it follows that a succession of thin walls isolated from each other by air spaces will be more effective, weight for weight, than one solid wall.

Soundproof boxes for noisy machinery are sometimes lined inside with a laminate of thin lead sheet and foam rubber, but if available space permits, then two wooden boxes one inside the other, connected only by rubber or cork, as shown in Fig. 9.5, will provide a high level of isolation provided each layer is airtight, with gasketed doors and labyrinth ventilation. Audiometric test rooms, which provide total isolation, are basically concrete boxes on springs inside outer concrete walls. Doors and windows also have two layers, each with independent sills and frames.

Ventilation might appear a problem in soundproofed rooms, but felt lined labyrinths and canvas coupled pipes attenuate sound traveling from outside, while maintaining the principle where no rigid coupling exists between inner and outer wall structures (Fig. 9.6). Another system which maintains air-flow but inhibits sound transmission is a variation on the labyrinth, making use of the isolation cavities of the walls.

Commonly three inches wide, part of the cavity is converted into a duct with polyurethane foam strips, which are sufficiently flexible to contain

Fig. 9.5 Isolation boxes connected by rubber blocks demonstrate the principle of acoustic isolation.

air pressure without acoustically coupling the walls. Entry at the top of one wall and out-flow at the bottom of the next, forms an effective sound isolated ventilation duct. Fig. 9.7 shows how the ducts can be placed inside an isolation wall consisting of three layers with two air-spaces. Pre-sealing the inside wall surfaces in the duct area prevents air contamination.

If two inch material is used for each outer layer with three inch air gaps on either side of a four inch, rock-wool insulated, plaster-board clad frame, then the whole assembly takes up no more than fifteen inches, yet has between 80 and 90 dB noise rejection at all except subsonic frequencies. Acoustic test reports on the various building materials are usually available from suppliers.

Lightweight dry construction was used to build the two theaterettes shown in Fig. 9.8 in which this technique was employed. The 22 foot shell of each room was a four inch frame wall built on an existing suspended concrete floor, and isolated from the original ceiling. The projection

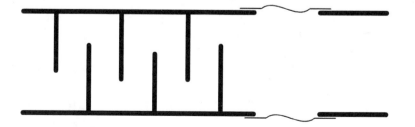

Fig. 9.6 Air duct labyrinth and canvas coupler.

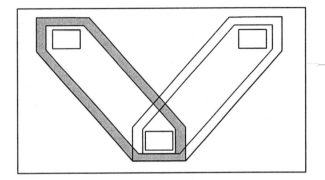

Fig. 9.7 Two-stage isolation wall with offset air-ducts.

room side was clad in two inch compacted and hardboard faced straw-board, and the other side with 5/8" plaster board. The whole frame filled with sawdust. Any wiring passing through the wall was contained in screwed steel conduit.

Independent floors of three quarter inch chip-board were laid over a layer of slack bitumen paper, on timber bearers supported by cork blocks. Two inch compacted straw-board was erected on the elevated floors to form air-gapped isolation walls. Convex "spoiler" ceilings of the same straw-board were supported on sheet-metal beams resting on the frame walls.

The announcer's booths had secondary independent walls, with saw cuts along the floor to isolate them. Solid core doors were fitted through-out, and gasketed with adhesive buffer tape all round. Air conditioning was piped into the theaters though a single inter-wall duct, and a double inter-wall duct into each announcer's booth as shown in Fig. 9.7. Return air left through a passive ceiling grille. The acoustic panels shown in Fig. 9.2 were glued and nailed on to the inner walls and painted.

Double quarter inch glazing was used for all windows, and although the isolation between booth and theater was limited due to the large win-dows, noises outside the theater never interfered. Thirty two interlocked film recorders, projectors, and playback dubbers, plus two transfer suites, operated immediately behind the single airspace wall, serving the two theaters shown and a larger one on the other side.

All types of voice recording was done in both theaters, including post-sync and perspective work. The isolation was such that a child's voice could be recorded in theatre two booth, while a 16 or 35 mm projector started and stopped five feet away on the theater three projector

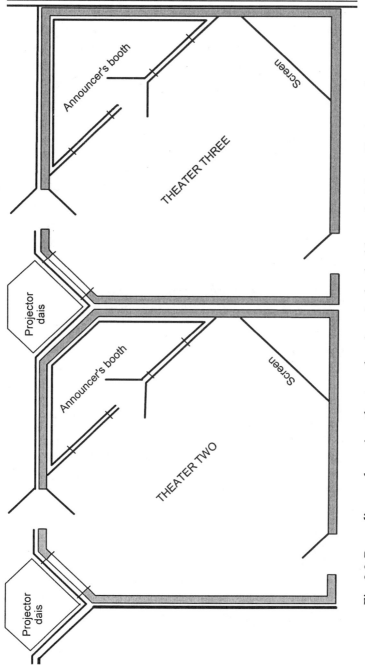

Fig. 9.8 Recording and preview theaters showing single, double, and triple wall construction.

dais. The dais was a timber riser eighteen inches high, built on top of the elevated machine room wiring floor.

A photograph of the machine room running the length of both theaters behind the projectors is shown in Fig. 9.9, indicating the need for good isolation. Theaters two and three are to the right of the picture. A third larger theater to the left of the machine room was served by three film projectors, making a total of seven. Add to all this the sound of voices raised above the rattle of machinery and a fairly good idea can be had of the airborne noise level.

Impact noise was dealt with by the raised wiring floor. Ceiling spaces are another source of airborne sound transmission, but their isolation is assisted by the relatively long path they provide. In building environments, airborne sound roughly follows the inverse square law in which a radiating body in free air will produce much the same reduction with distance as light or radio transmission:

SOUND TRANSMISSION IN AIR DECREASES BY THE SQUARE OR THE DISTANCE FROM THE SOURCE

OVERCOMING BUILDING NOISES

A compact film studio was built in Hong Kong on the twenty-second floor of an office complex. The dubbing and preview theatre, shown in Fig. 9.10, had fluorescent lighting, but the tube ballasts and starters were installed outside in the foyer to keep the theatre noise level to a minimum.

It is difficult to perform an accurate noise survey before an area has been enclosed, and when the small announcer's booth was finished, it was found that the elevator motors for floors 1 to 18, which were on the same floor as the theater, produced a broad spectrum noise transmitted by the concrete building that was impossible to remove with a high pass filter. Independent walls and floor walls were built inside the booth with angle iron, isolated from the original walls, and supported by cork blocks. The wall frames were clad with 2 mm lead sheet and the ceiling fixed on top after dressing the walls with pleated fabric.

The isolation achieved was in every way satisfactory, but the booth was so dead that curved laminated plastic panels on frames had to be fitted on adjacent inside walls to brighten it up.

This demonstrates the efficacy of very short reverberation times for putting the bite back into voices. More distant reflectors can give a "bathroom" sound, but this is not evident when the reflector is one or

Fig. 9.9 Machine and projection room for three dubbing theaters.

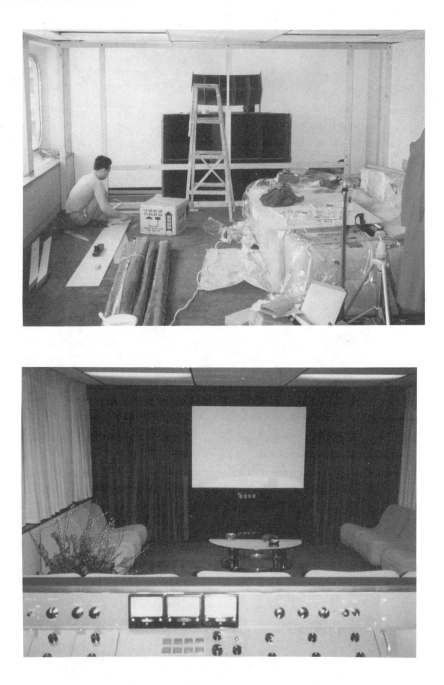

Fig. 9.10 Hong Kong dubbing theater during construction
and after completion.

two feet away. Work requiring a dead studio, with reverb to be added back later to balance recordings with others, is easy to achieve by simply moving the reflectors out.

Announcers booths are not a perfect way to record voices, but with a little ingenuity they can be quite workable. The best all purpose voice and post sync. recording facility is an acoustically dead theater with provision to move hard and soft baffles around, the hardware to create sound effects, and enough space to simulate most acoustic environments. But at various times and places, the principles of acoustic isolation and control can be relied on to get the job done under less than ideal conditions.

PLANNING AND SUPERVISING A BUILDING PROJECT

Most builders are unfamiliar with both the need and procedures for building isolation rooms. There are two choices. One is to find a builder who has done it before, and the other is to coach a new builder in the technique.

Here is a short list of procedures you will need to convey to your construction team:

1. Isolate the frame from other noise bearing structures, using cork or rubber. Because of the weight of frame members, fairly dense compliant materials will be needed, like Vibration Damping Cork bonded with rubber, which is made for the purpose of isolating heavy machinery from concrete floors, or sagging will occur and isolation will be compromised at some future date. Isolation walls that drop and touch other parts of the building can not be fixed.
2. Preserve air-tight conditions, especially at concealed non-obvious places. Stuffing window and door frames with newspaper before the architraves are fitted works, although there are materials available that do it better. Caulking around architraves with suitable compound is good, but it is better if the cavities between door and window frames and the wall are filled. Filling door frames with sand works very well, but it tends to run from the smallest hole after it dries.
3. Make air-gaps between isolation walls wide enough to accommodate material warping or frame movement. Saw-cuts separating the halves of double glazed windows and double doors can be concealed with felt strips, but avoid having air blowing through double window cavities, to keep the glass clean longer.

4. Choose sound absorbent and wall packing materials carefully. They not only have to be efficient, but they also need to be fireproof, easy to clean, and paintable in some cases.

5. Make sure the ventilation system is geared to operating at a low rate of flow; standard air conditioning won't do it. Buildings radiate heat inward in warm conditions, and absorb heat in cool conditions. If air flow is low enough to be silent, it will definitely need to be heated or cooled, to maintain comfort.

6. Fit passive vents to every room, with a chicane, or felt lined zigzag construction to prevent noise transmission. The air pressure differential in a sealed room with even a small fan blowing in or out will cause the walls to collapse if the pressure can not equalize. Imagine a pressure of only one pound per square inch multiplied by the total wall area in square inches; it may total thousands of pounds.

7. Compliant panels acting as absorber resonators are contrary to the principles of every self-respecting carpenter, who likes to firm everything in place with plenty of framing. Take time to explain how they work, and that they must be able to "wobble."

Whichever way it is done, your intentions and details of the methods employed to achieve them will have to be extensively documented, with detail drawings or sketches, notes on procedures and the reasons for them, together with a lot of patient liaison.

It's not easy to ask experienced builders to learn new techniques, but in an age when it is usual to purchase packaged technology for instant use, many people are more than happy to have the chance to work on something quite different.

SILENCING A GENERATOR

There are many effectively silent machines in use, from air compressors to mobile generators. In common, they have the attributes of breathing air, the need for cooling, and being inherently noisy. The technology that makes them quiet is old, but the information about it is freely available.

The following events took place at a film production studio a few decades ago when much of the sound and camera equipment required 220 volt three phase power. That's not the way most of it works now, but the principle of soundproofing an engine still applies to transportable lighting generators and utility alternators.

The point in outlining the facts is that not everyone is familiar with a particular principle of physics, and need to be launched on the right track where air-flow and air pressure are concerned. So if you are one of those fortunate people to whom everything is straightforward and obvious, then this discourse is not for you.

A portable alternator driven by a two cylinder gasoline engine had been acquired for a particular job, and as is often the case, it was put straight to work at the end of a long cable because there was no time to soundproof it. When time was available to prepare it before the next production, a trailer was purchased, and a padded box (Fig. 9.11a) constructed to an engineer's plan. No instruction was given to do anything except bolt it in and close the lid.

The alternator set had a mid-mounted fan, and the people who installed it on the trailer were confident; "Well, the fan blows from one end to the other, so it will look after itself." Before the test, it had been suggested that a baffle would be required around the fan, to separate the two sides, like a speaker baffle. But as is often the case, free advice is not as welcome as one's own.

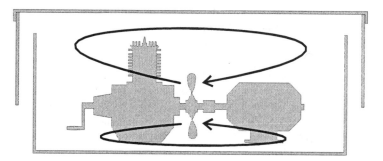

Fig. 9.11a Air turbulence in the box without baffle.

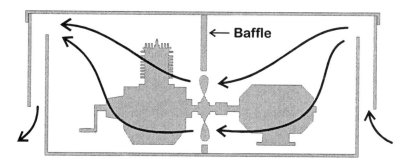

Fig. 9.11b Airflow through baffled fan.

The alternator in its box was certainly very quiet, but after a ten minute test-run, it started to slow down. The cover was lifted off to reveal a cloud of smoke and a heat stressed engine. The baffle was duly fitted and the alternator re-started for a successful test run (Fig. 9.11b). Half a sheet of paper was torn into small pieces in the manner of an initiation ceremony, and thrown into the low pressure end of the alternator box. Within two seconds, all the pieces shot out of the exhaust end of the box.

Just as speakers and electric motors have things in common, so also speakers cones and engine ventilation systems work by the same rules. In physics, there is no *Suck,* only *Pressure.* Vacuum is absence of pressure. A fan works because it moves air from one side to the other. The high pressure end of the alternator box releases air into lower pressure atmosphere, but the low pressure end does nothing. It is atmospheric pressure at 14.7 pounds per square inch that pushes its way in to relieve the pressure differential.

Standing beside the trailer revealed a low rumble. Ten feet away, the alternator was for all intents and purposes inaudible. It performed years of service, running for four hours on a tank of fuel, two fills per day (stop the engine before filling, hot exhaust particles are guaranteed to ignite the fuel-air mix displaced from the tank).

It is difficult to completely soundproof a machine or a room; even the action of closing a door in another part of the building makes a noise that will pass through walls. There will be an acceptable frequency below which subsonics pass almost unattenuated, but sound equipment conveniently bandpasses itself in the majority of cases, and where it does not, convention is that the inaudible subsonics are not admitted to any serious extent, either by normal dialog bass roll-off, or by purpose filtering.

Chapter 10.

Audio Cabling

Running cables is one of the most regular assignments of the audio technician, whether in a portable system or a fixed installation. The techniques vary according to cable length; single shielded can be used for short lengths in domestic systems, but twin shielded cable is used for longer runs, and all professional installations.

Twin shielded cabling is called for in larger domestic installations to provide satisfactory noise immunity, but the method of connection is critical to achieving this result.

The purpose of domestic and professional audio is essentially the same, although multi-generation work and long lines have to be of higher specification to achieve it; an end result with a frequency response of 20 to 20,000 Hz or better, and a signal to noise ratio of 70 dB or better. Larger domestic installations often run at line level into several rooms, and a purchaser who has just spent over $1,000 will not be happy if the system has audible line noise or poor frequency response, any more than will a radio or television broadcaster.

BALANCED AND UNBALANCED LINES

There's often a lot of talk about the merits of both methods of handling signal at line level. There is also a great deal of confusion as to exactly what constitutes a balanced or an unbalanced line, input, output, or system, and why either one is necessary. But the most important factor remains the noise immunity of audio lines under various conditions. Here are some explanations.

Balance does not simply mean both sides of a circuit are above ground. Balance can exist to a useful degree even when the signal neutral is grounded. The main question is: "Balanced to what, and why?"

All signal lines, whether long or short, high or low impedance, or inside equipment racks tied in looms with other cables, are subject to induction of spurious signals or noise, which once mixed with the program can not be extracted. Even before shielding, there is a technique which is applied to minimize the introduction of noise; by making the line, or positive side of the signal pair, and the neutral, negative, or grounded side, physically identical, with the conductors twisted together. In this way, anything external that happens to one conductor, occurs equally in the other, but out of phase as far as the signal is concerned; so induced noise cancels.

This explains the reason why single shielded cable, with the exception of coaxial transmission lines which work in a completely different manner (Chapter 3), is limited to short runs, like stereo inter-connect cables and point to point hook-up within equipment modules, while twin shielded cable is used for all lines of any significant length, wiring within walls, all multiple cabling within a rack, and lines that unavoidably lie near power wiring, which is a prime source of mains frequency harmonics and RF switching transients.

While fully balanced lines, inputs, or outputs, have both sides above ground, it is still necessary to apply all the principles of balance to these lines, just as with lines with a grounded neutral. Provided the lines are not too long, and other conditions are in place, like freedom from power mains ground loops, lines with grounded neutrals can be just as noise immune as any other.

Both grounded and floating neutral lines need to be EFFECTIVELY BALANCED to have high noise immunity.

In the past, professional audio modules had transformers at input and output to provide both balance and impedance matching. This was necessary when vacuum tubes used high voltages that had to be isolated, and operated in high impedance circuits from 5,000 to 100,000 ohms, incompatible with line impedance, but today solid state input and output stages can work directly to line, although some outputs use a low ratio output transformer.

There is nothing wrong with using unbalanced lines, provided the cable is connected in a noise immune configuration described in the following pages and shown in Fig. 10.1b. General practice is to use these techniques:

1. Twisted pair line and neutral program conductors.
2. Low impedance line driver signal sources.
3. A close woven or foil plus drain-wire shield, connected to neutral **at one end only.**

4. Avoidance of noise loops by making sure that there is only one neutral path, and that the shield does not carry signal currents.
5. Add-on 1 : 1 transformers where noise problems are encountered.
6. Either unbalanced inputs or electronically balanced inputs, in which line and neutral connections are fed directly into the non-inverting and the inverting inputs of the first stage; an operational amplifier, either IC (integrated circuit) or discreet components.

Electronically balanced inputs should be balanced up to RF, because radio frequency noise bands can also cause audio noise (pages 211 and 212).

The illustrations of Fig. 10.1 compare the different types of cable used for short and long runs. Use of single shielded cable is confined to audio wiring inside equipment modules, and for interconnecting domestic audio equipment. If single shielded cable has become identified with domestic Hi-Fi, it is only because these systems usually do not normally extend at line level beyond the confines of one room. It is essential to grasp that the aims of professional audio system installations are no different; it's just that they are usually more extensive, and exist in complex multi-mode interconnections that insist on twin shielded cabling because of the complexity of earthing, or grounding, conditions, and the greater length of the cable runs.

Fig. 10.1a Single shielded cable and connections. Noise immunity is poor in long cable runs.

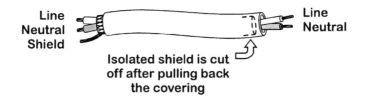

Fig. 10.1b Twin shielded cable and connections for unbalanced circuits. Neutral and shield are strapped together *at one end only* so that noise and signal currents do not share the same conductor.

There is always noise; it just has to be kept at arm's length. A line can be effectively balanced even if one side is grounded. Even fully floating lines need to obey the basic rules to keep them noise immune. When trouble strikes, both noise due to unbalanced conditions, and noise due to mains ground loops, can be fixed by installing one to one line isolating transformers at the destination end, restoring or creating line balance; but transformers have to be shielded, mounted or oriented away from hum fields like power transformers, and suitable in frequency response and signal handling capability for the line level in use.

SHORT RUN AUDIO WIRING

The illustrations of Fig 10.1 show the two main types of shielded cable. Lines up to 50 feet or 15 meters long can be run in single shielded cable **provided it is driven by a suitable low impedance source**. To this end, output amplifiers of various modules are called **Line Drivers**.

When in doubt about the capability of the line driver stage, divide the suggested maximum length by five, or by ten if mains cables are present. It's not just the signal voltage level that governs noise immunity, but also the impedance of the source, because if the line impedance is relatively low, it short circuits the noise signal. Separate shielded cables should always be used for left and right channels to minimize cross-talk. All line level wiring should be shielded, assuming normal line driver output voltage and source impedance.

LONG BALANCED AND UNBALANCED LINES

In longer lines, **twin shielded** cable is necessary, primarily so that leakage currents in the shield do not share the same conductor with program neutral. Identical line and neutral conductors inside a floating shield (connected one end only) therefore offer better noise immunity than a single line conductor in a common shield. A shielded cable has to reject more than just mains hum. Transient noise in most buildings comes not only from local equipment switching on and off, but also from more distant sources. RF hash is always present in the power mains by direct injection into equipment, and also by radiation from wiring, which means that using a line filter to supply equipment does not guarantee freedom from noise. The noise spectrum up to several megahertz has to be catered for, because non-linear stages that exist to some degree in any audio equipment will detect the RF noise signal and reproduce it as audio.

If line level cables run into other rooms, then twin shielded cable is necessary. The shield is connected at one end only, unless required to earth-link two other cables or equipment modules. There should be only one earth or neutral path in unbalanced systems. Wiring done in this way makes the program conductors relatively independent of noise currents. In complex cable installations, a third conductor, called a **Technical Earth,** is sometimes used for the purpose of earth-linking to avoid passing total noise current through the full shield lengths. Microphone extension cables using three conductors inside a shield, for example, connect each shield length to the continuous ground wire at the destination end only, resulting in high noise rejection where cables have to share floor space with power and data cables in television and motion picture stages.

When laying audio cables it is inevitable that they will pass near or cross power wiring. A distance of at least three feet should separate power and audio runs, although they can be run closer for short distances if it is unavoidable, and if it is necessary to cross power lines, it should be done at right angles.

Occasionally, the situation will arise that various audio modules are supplied from different power outlets, and that due to ground leakages in other electrical equipment on the same power circuit, there exists a small potential difference between the earth connection on each power outlet. This can result in alternating current flowing between equipment modules through the shields or neutrals unless precautions are taken. There are two courses of action. For safety reasons the earth wire on any equipment can not be disconnected, but disconnecting one end of the program neutral line, as shown in Fig. 10.2, may break the "earth loop" or "hum loop," and program neutral will have continuity through the mains earth wiring only. The open circuit neutral remains connected at the origin end to serve as a balance for induced RF voltages.

Severing a program neutral is not a valid design intention when planning an installation, but in many cases it will get the system working until alternative arrangements can be made. However, if the noise originates in the source unit, due perhaps to leakage from a power transformer or a de-laminated circuit board, then this method will not help, and it will be necessary to resort to the correct way to solve the problem; use of an audio isolation transformer. An audio isolation transformer of 1 : 1 impedance ratio fitted at either end of the line positively resolves the problem by breaking the leakage path (Fig. 10.3). A transformer at both

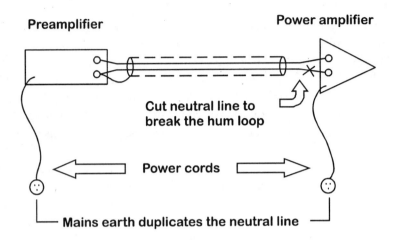

Preamplifier

Power amplifier

**Cut neutral line to
break the hum loop**

Power cords

Mains earth duplicates the neutral line

Fig. 10.2 Cutting the program neutral at one end may solve a noise problen due to an earth loop.

ends would create a fully balanced line, but this probably won't be necessary. The transformer is commonly 10k to 10k ohms, and has the ability to reflect load impedances, including capacity and inductance, both forwards and back to the source.

Connecting audio isolating, or balancing, transformers into lines should not be overdone. They are expensive, they constitute a certain quality loss no matter how good they are, they are inductive, may resonate with stray capacities at nuisance frequencies, and they may pick up noise from mains wiring or power transformers, because even though they are shielded, shielding is only relative.

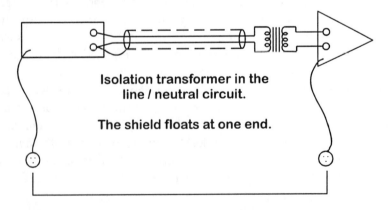

**Isolation transformer in the
line / neutral circuit.**

The shield floats at one end.

Fig. 10.3 An audio transformer breaks the noise loop.

Audio transformers work best with a low impedance source; the lower the better, but they can't operate correctly if the load that follows is lower than their specified impedance, because they are asked to pass too much current and the core will saturate, reaching its maximum magnetization, and so it will not transfer a true waveform. The transformer will need to be of suitable frequency response and noise rejection, and capable of handling line level and its program peaks. A microphone transformer won't do the job because it can't handle line level, which will cause frequency response errors and distortion due to core saturation, but shielded 10,000 ohm line level audio isolation transformers as small as an inch cube are available from broadcast equipment suppliers. 600 ohm transformers are larger as they are handling more power even though the signal voltage is similar. Two transformers are required for a stereo signal, and they should be mounted far from power transformers or oriented for minimum noise pick-up.

Transformers have to be connected in phase because while unbalanced wiring can only go one way round, any balancing device gives one the option of inverting the signal phase. While talking about "phase inversion," it has to be said that audio modules other than portable radios and the low end of the audio market are by design **Phase Coherent**, in that the same portion of a positive going signal at the input will be positive going at the output.

This is particularly important where various sources are to be mixed, because the first wavefront of most sounds is higher than the following wave peaks, and if the available head-room of the transmission media that follow—recorders, broadcast transmitters, and so on—is to be used to best advantage, then all signals from microphone to speaker should have their first wavefront on the positive side of the signal waveform.

Sound is not symmetrical. The importance of phase coherence is that phase should not be arbitrarily reversed, or inverted, when installing lines, transformers, and the like. Doing so may increase the dynamic range of the program, resulting in poorer effective signal to noise ratio, and loss of impact and realism. Speakers should produce positive air pressure on initial wavefronts, in the same way that microphone diaphragms move inward when sensing positive pressure.

REDUCING NOISE INTERFERENCE AT SOURCE

Regardless of efforts to keep noise out of audio and other systems, there are persistent cases where switching transients from the power mains just won't go away. Once the noise signal has been radiated by the power wiring, mains filters have little effect, and the best efforts to immunize systems by following all the rules of wiring and grounding are to no avail.

In these cases, suppressing the noise at its source is often the only way out. To understand the mechanism by which power mains glitches are produced, it's necessary to refer to refer to Chapter 1, and review the passage on **Back-EMF** generation. Whenever a switch opens, interrupting current flow, the inductance of the circuit converts the energy from the collapsing magnetic field into a high voltage transient called the Back-EMF, which is higher than the original voltage because the inductor circuit becomes very high impedance once the switch is open. Everything has inductance, even a light circuit, and unless a **zero-crossing** device is used, breaking the circuit at the zero current part of the AC waveform, the incidence of switch-off transients in a place like a recording studio means that every motor, light circuit, or transformer, may have to be fitted with some simple device which will reduce its switch-off transient at least 50 dB.

Many people have used capacitors connected across the switch contacts, but this is not a particularly effective method, and it can be an electric shock hazard since the capacitor still conducts AC even though the switch is off. A more effective and completely safe method, requiring smaller capacitors, is the spike suppression network shown in Fig. 10.4. It does nothing for switch-on transients, but these are usually very small compared to the enormous glitch that can occur when breaking a circuit.

The value of the components varies with supply voltage. A 100 ohm half watt resistor works for all mains voltages, but C is 0.22 mfd for 120 volts and 0.1 mfd for higher voltages.

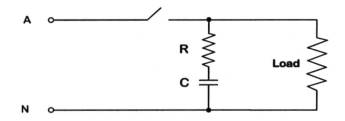

Fig. 10.4 Spike suppressor circuit.

The peak voltage rating and type of the capacitor is important, as it is dealing with Back-EMF peaks well in excess of the supply voltage, but the current involved is very low. For voltages up to 250, a peak of 630 volts should be catered for. In proportion, higher voltages require 1,000 or 1,500 volt ratings. Polyester capacitors are suitable for this application if enclosed, but components rated for the mains voltage in use are prefered.

SAFETY

Anything made for direct connection to the mains needs two features in its construction, and a small device like a spike suppressor is no exception. It must have a case around it, unless it is enclosed inside existing equipment, and it must be **fail safe**. This means that it must go open circuit before catching fire or exploding. The spike suppressor should therefore be made with a low wattage resistor fragile enough to fail instantly in the event of an overload. If the unit is encapsulated, it should be in non-flammable material. Safety components are discussed in Chapter 11.

The suppressor is available ready made from a number of electrical component manufacturers, and can be seen in many appliances ranging from photocopiers to refrigerators. As with many devices, it may be cheaper, and probably safer, to use a commercial one rather than make it yourself.

Connecting the suppressor across the switch is unsafe and less effective. Devices across switches maintain the power partly on when it is expected to be off. The suppressor should go across the load, so that when the switch is in the off position, the device is right out of circuit, and neither it or the load passes any current or has voltage across it.

There are other methods of suppressing surges. One is to use a varistor to limit instantaneous line voltage. and discriminate against spikes. These devices, available also in power outlet protection units, offer some security against power surges caused by lightning strike and supply switching errors, and are cheap insurance for equipment.

Direct current circuits can be de-glitched by connecting a diode across the load, in opposition to the supply polarity. A low voltage relay can generate just as big a noise problem as a mains circuit because as explained in Chapter 1, breaking inductive circuits generates Back-EMF voltage, causing the transient. The inverse diode is very effective in quenching the Back-EMF surge in DC circuits (Fig. 10.5), although the RC device can also be used.

Inductors produce voltage with very little current when their magnetic field collapses, so a small current diode will suffice, but the voltage rating should be at least 100 volts, or five times the relay supply voltage. The RC suppressor of Fig. 10.4 is effective on low voltage AC relay coils.

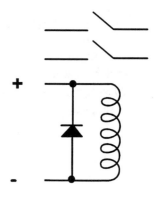

CONNECTIONS TO VARIOUS TYPES OF EQUIPMENT

Fig. 10.5 Diode spike suppressor.

Balanced equipment connection is straightforward, as the **Line**, **Neutral**, and **Earth** terminals are the only ones available, and are not normally linked. But unbalanced equipment can specify a variety of input and output connections due to differences in internal circuitry. Fig. 10.6 shows the alternate connections that may present themselves,

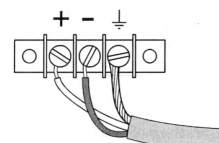

Conventional connctions of balanced line cable.

The same connections may suit unbalanced lines, where shield or neutral are either not connected, or they are tied together at the far end.

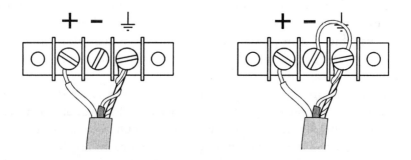

Fig. 10.6 Illustrating conventional (top) *Barrier Strip* connections, and the variations that may be specified in manuals for unbalanced connections at the manufacturer's discretion.

demonstrating the necessity of reading the Owner Manual when installing any equipment that is not completely familiar. The owner manual will indicate the method, which may be different for inputs and outputs. The connection variations illustrated may also apply to XLR connectors. Incorrect connections may exhibit confusing faults, like the inverse curve from an active crossover.

Owner Manuals are the most important accessory of any piece of equipment. Whenever equipment is installed or supplied, it is customary to pass the Owner Manual to the customer together with the Block Diagrams of the system, and any information that a future service technician might need. Keep a copy for yourself too so that you can give the customer informative advice by phone if difficulties occur.

There is another trap that can be experienced when connecting the input of some stereo equipment to balanced lines. In the case of the amplifier arrangement shown above, strapping both source neutrals together will solve the problem; a situation that does not occur with unbalanced sources, as they invariably have both neutrals connected at one end or the other. Many amplifiers expect to be driven by an unbalanced source. If the system of Fig. 10.7 was operated without the strap, the two amplifier inputs would have have no output from one channel, or one channel would be out of phase.

Again, it is necessary to stress that stereo equipment is not two channel equipment. Stereo systems take short-cuts in grounding and channel separation that have no adverse effect on results provided they are used as

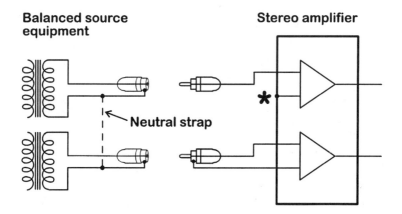

Fig 10.7 **The diagram shows how some stereo amplifiers do not carry the neutral connection through to both input sockets, but ground one to chassis instead.**

what they are; left and right channels of an integrated system. In a two channel system, Channel 1 and Channel 2 are independent under all conditions.

CABLE SEGREGATION AND RACK TERMINATION

All cable groups representing different functions should be kept separate as far as possible, to minimize interaction. Depending on the specification for the installation, cables may have to be fully isolated in steel conduits outside the rack. To couple such a system to the rack, flexible steel conduit in various sizes can be coupled to holes cut in the side of the rack, or preferably brought in through the floor of the rack via a wooden riser three or four inches high. Rack systems and general cabling are covered in Chapter 14.

Large clearance holes, or a completely open rack floor enables flexible conduits or cable bundles to be secured to the tie-bars up to the level of the first horizontal cable deployment. Cable segregation, like choice of conductor size, depends very much on the distance that different cable groups will be running together. There are instances where it is possible to get away with running power and audio together for short distances, as in multi-function patchcords, but the need to give noise immunity its best chance should be at the front of every cabling decision, especially where low level cables are concerned.

THERE-AND-BACK CABLE TERMINATION

The "There-and-Back" method is shown in Fig. 10.9, and suggested as an alternative to Fig. 10.8. Cables are run past their termination point, or to the top of the rack, and then back to the the equipment they serve. This is a method which uses a little more cable, but pays off where it is likely that rack-mounted modules may be re-positioned later, whether it is during the installation or at a future date. Many installers like the neat appearance of cable running tightly to the terminals, but mistakes can't be corrected, the ability to change anything later is lost.

This method of bringing cables to a rack applies particularly to jack strips, where a securely tied cable-form, with two inch tails going to each jack, provides mutual mechanical strength to take the strain off the conductors, but gives complete flexibility for change. A patchbay cannot

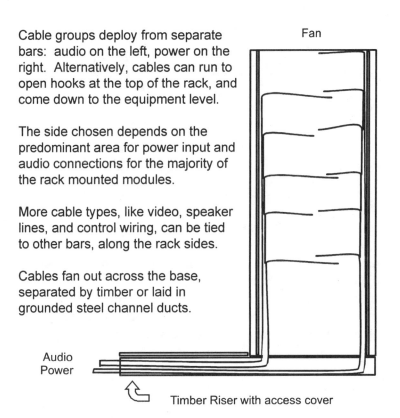

Cable groups deploy from separate bars: audio on the left, power on the right. Alternatively, cables can run to open hooks at the top of the rack, and come down to the equipment level.

The side chosen depends on the predominant area for power input and audio connections for the majority of the rack mounted modules.

More cable types, like video, speaker lines, and control wiring, can be tied to other bars, along the rack sides.

Cables fan out across the base, separated by timber or laid in grounded steel channel ducts.

Fan

Audio
Power

Timber Riser with access cover

Fig. 10.8 **Rack cabling seen from the back.**

be easily disconnected if it becomes necessary to change plain jacks for self-normalling, for example, as discussed in Chapter 14, or if jack positions need to be changed. If the wiring is inaccessible between adjacent equipment, leave a free loop of cable at the entry side of jack-strips, so that the strip can be swung out front for access.

The nineteen inch rack is so dimensioned that most people can't get their head and shoulders inside and still work, so soldering to jack strips is best done at the front. If anything is going to be altered after an installation is completed, it will be to add or change the position of jacks on the line level patchbay. It's a good practice to make jack strips accessible without having to remove adjacent equipment.

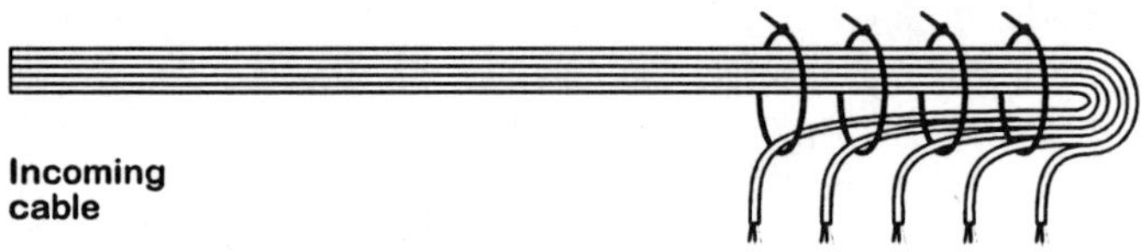

Fig. 10.9 Method of terminating cable on a jack-strip.

PRE-MEASURED CABLE RUNS AND JUNCTION BOXES

Sometimes it's necessary to work to measurement when running cable, especially where a bundle has to be pulled through a pipe or a wall. In long runs, therefore, allow an extra 6 feet at each end if the cable has to be cut to approximate length before it is pulled through its path. The spare pieces left after termination will always be used to make patchcords; if not in the current installation, then in the next, and the tails will not be wasted.

There arises a situation based on economy and convenience where cables change type at one or both ends of their run. An example would be speaker lines, where electrical grade cable is ideal, but is not sufficiently compact or flexible to connect directly to the amplifiers or speakers.

In these cases, the non-flexible type cable is terminated in junction boxes, and single or double insulated flexible cable is run the rest of the way. In these cases, particular care should be taken to ensure that the junction box connections have long term reliability. Speaker and power terminals get hot when loose, and this makes them looser still. To avoid an avalanche effect leading to failure, follow-up maintenance on screw terminals includes re-tightening at least once, because the copper conductor shrinks and work-hardens under the pressure.

If junction boxes are used, paste a chart of the cable number, position, and function, inside the cover. Even the installer won't remember what the connections were two years later, and installations are very much an information based business.

PREFABRICATED CABLE LOOMS

Cables running to all parts of a compact system, like a recording studio, can often be prefabricated looms which are made to measurement from a floor plan. The tied bundles of cable are then dropped into wiring ducts and connected at every position.

The cut length of each conductor has to include not only the floor length, but also the rising height of the equipment terminations. As with any lengths cut to measurement, it is necessary to allow extra for small errors, and the there-and-back method of terminating cable ends should always be used for this type if installation.

Pre-looming can be a great advantage where the site working time is at a premium. The Hong Kong recording studio shown in Fig. 9.10 was scheduled to be handed over to the owners six weeks from the day the partition walls were started. The floor plan was drawn on a seven day survey trip, the equipment made ready at home base, and the whole system transported by container. No cables were forgotten because a Cable Schedule had been prepared from a block diagram of the whole system.

Preparation was a matter of laying out the floor plan on an available area, with nails in the floor indicating every turn and cable end and, as usual, measurements were double checked before cutting any cable. Power systems were terminated and test-run before despatching. It could just as easily have been done outdoors with six inch nails or tent pegs.

On another installation far from home, the cable was pre-loomed around weights in an adjacent corridor while the site was still under construction. In this case, it was not possible to spread it out like the floor plan, but the turns were all made at the right places and only the angles modified to accommodate the width of the working space. It is often necessary to work around builders, electricians, and carpet layers. In this case in particular, pre-looming the cables avoided the need to work surrounded by ladders and concrete dust, and all the usual hassles when there are too many people working in very small rooms.

The time factor is a little different to an ordinary installation. Overall, pre-looming might be seen to take a little longer, but working time at home base is cheaper than site time, and twice as convenient, since a completed loom can be dropped into place after all other construction teams have finished, and the site is clean.

KEEPING RF OUT OF SYSTEMS AND CABLES

There are some circumstances that really try the skill of audio installers. I refer to unacceptable noises that just won't go away. Most are outside our control, and some are intermittent. They are Radio Frequency Interference, or RFI, caused by motor commutators, square wave power supplies, thermostats, welders, projection lamp igniters, fluorescent tube starters, lighting dimmers, and a whole range of local and distant power glitches.

Radio frequencies admitted to an amplifying system are detected by any minor non-linearities in low-level stages, and appear as audible noise. RF will find its way through the most insignificant path, and the measures to combat it can be compared to making soundproof rooms airtight by blocking every possible way. Methods include using a series RF choke followed by a shunt capacitor to ground at susceptible inputs, and in equipment using discreet transistors, a small value capacitor connected between emitter and base of the first stage blocks RF without affecting the audio band response.

One of the most effective ways to stop RF getting into microphone extension cables is to ground all connector shells to the shield. However, a preamplifier input may have up to four connections; line, neutral, shield, and chassis, and the connector body or shell automatically makes contact with the chassis unless it is an insulated type.

It is undesirable to connect the shield to the connector shell where the last microphone cable in the line goes to a mixer or preamplifier input as this can introduce noise loops by inter-connecting system ground and chassis ground. The method shown in Fig. 10.10 avoids this without ambiguity while maintaining mutual RF protection for all line connector pairs coupled together, right back to the microphone.

There is a further hazard associated with grounding the connector shell at the microphone end. Not all connectors or microphones make perfect contact at the outside of the shell. When contact is poor, making and breaking erratically, it will produce all the noises imaginable. So the advice to ground microphone end cable connector shells needs to be taken

Connector shell facing microphone is grounded to the shield

Amplifier end connector shell is not grounded inside the cable

Fig. 10.10 Grounding origin end only connector shells of microphone extension cables rejects RF and avoids noise loops.

with some reservation. Many technicians prefer not to ground them until faced with RF problems.

GROUND TO THE SHIELD ONLY THE CONNECTOR SHELLS AT THE MICROPHONE END OF EXTENSION CABLES.

One of the most difficult interference noises to treat is a narrow beam Radar signal, which sometimes identifies itself as a sound like a bee flying past your ear. Airports, defense installations, and shipping, all sweep high intensity pulsed microwave beams in a regular pattern. While the duty cycle is negligible as far as dangerous radiation is concerned, the instantaneous intensity can be very high, and as usual with extreme frequencies, its effect is quite unpredictable; it may do nothing at all.

If you are bothered by this bane of audio technicians, I suggest asking the people making the noise how they handle their own RFI, because there are sure to be sensitive inputs in their own domain requiring some specific type of protection.

Chapter 11.

Transformers and
Power Supplies

TRANSFORMERS

Voltage dividers are networks that can reduce voltages and signal levels, but they they do it by wasting a high percentage of the power, and they can not multiply voltages. Transformers are among the most efficient devices, and can divide and multiply voltage, current, and impedance. Only the small inefficiency factor of the transformer is wasted in heat, second only to the losses occasioned by speaker line, crossover network, and voice coil resistances.

As demonstrated by the Constant Voltage Line transformer shown in

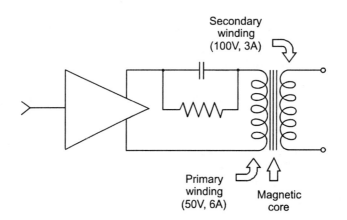

Fig. 11.1 An amplifier to line transformer converts voltage and impedance at 300 watts, demonstrating that mains power and audio are the same type of energy.

Fig. 3.3, and shown again in more detail in Fig. 11.1, audio signals and AC power are made of the same stuff.

A 300 watt amplifier delivers 50 volts across an eight ohm load, to take a round figure example. 50 volts at 300 watts equals 6 amps. To step up to a 100 volt line we use a transformer turns ratio of two to one; as voltage rises, current falls, and vice-versa, so at 100 volts, the current is 3 amps. Like a gearbox, a transformer's equivalent to Torque and RPM are also inverse.

Transformers convert primary winding current into a magnetic flux in the core, and back again in the secondary winding (Fig. 11.2). They have the ability to change the voltage, current, and impedance ratios, and since the intermediate energy form is magnetic, not electric, transformers can be made to fully isolate the direct current components of two circuits. AC can pass through a transformer, but DC will not, because generating a secondary voltage depends on the rise and fall, or reversal, of the magnetic flux in the core. However, the lower frequency components of DC pulses can pass in a form modified by the transformer's response.

An audio transformer's frequency response depends on many factors, which include the type of magnetic core material and the proximity, or "inter-leaving" of the primary and secondary windings (Fig. 11.6), since a proportion of the higher frequencies are transferred by magnetic induction directly from primary to secondary, bypassing the core.

This direct method of transfer is relatively efficient at very high frequencies, so that **electrostatic shields** are wound between primary and secondary in some power transformers to block noise voltages. The shield winding is grounded at one end only, so that it can not form a short circuit around the core. The same basic rules apply both to audio and power transformers, and since this book is dedicated to showing that audio and power technology uses the same principles and energies, the subjects of audio and power transformers are presented together.

As described in Chapter 1, all current carrying conductors have a magnetic field around them. The field can be concentrated by coiling the conductor so that the fields add, and further concentrated by placing a readily magnetized material inside the coil, so that the rise, fall, and polarity alternation of the current will charge, discharge, and reverse the polarity of the core. Electric energy in a transformer changes into magnetic flux (energy) and back again.

The core is a closed circuit to make it as efficient a magnetic coupling as possible, although there are exceptions where a gap-core is used as a current control method. With the exception of high frequency and

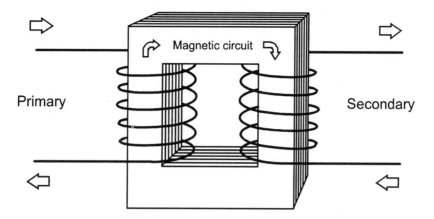

Fig.11.2 Basic transformer, showing windings and laminated core. Note that the direction of the coils determines their phase

switching power supply transformers using various grades of ferrite compound, transformer cores are laminated to prevent eddy current loss. A solid core would be seen by the magnetic field as a short circuit turn and it would pass large wasteful currents and get hot. Laminations are insulated from each other by a layer of surface oxide or lacquer. Anything that would form a short circuit turn around the core is insulated with a fiber or neoprene washer, like mounting bolts and the screws clamping the laminations. Typical core types are shown in Fig. 11.3.

Transformer core materials are specific to the frequency involved, various grades of silicon steel, mu-metal and radio-metal being some of the cores used for audio, while ferrite cores are used for RF and switch mode power supplies, which operate at high frequencies, and therefore can use a

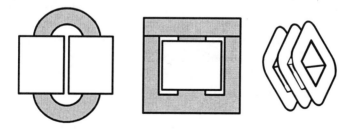

Fig. 11.3 C-core, wound from strip and cut in half for winding assembly, E-core lamination stack with layered insulation, and a nylon bobbin for side by side windings.

much smaller core. Transformers for low mains frequencies need larger cores, but less exotic core material, usually four percent silicon steel.

In the early part of the 20th century, mild steel laminations were used, and transformers were big and heavy. Metal laminations have to be annealed, or heated to redness, and slowly cooled, as in glass manufacture, and while mild steel is cheap, it eventually work hardens with continual magnetic fluxing, and becomes inefficient over the course of a few years, giving it in effect a "use by" date. Silicon steel is one of the most popular core materials for mains frequencies, as its efficiency does not deteriorate, and is only a quarter the weight of its equivalent in a mild steel transformer core. Silicon steel has a crystalline structure in which silicon boundaries form a cube lattice which partially insulates adjacent iron particles, reducing eddy current losses.

The most efficient power transformer is the toroidal type, with windings all the way round a ring-core made with continuous-wound strip, but the cost is higher because it needs specialized insulating materials and through-the-center winding techniques. Next in efficiency is the 'C'-core transformer, a strip-wound core cut in half and designed to be assembled into a pair of pre-formed balanced windings. The strip wound core transformers have a small external field, and the balanced winding formats, like toroids, and the two opposed primary-secondary assemblies on the cut and reassembled c-core transformer, match the field integrity of their cores. However, the E - I lamination core transformer has not been bettered for economy and utility, and their use outweighs other types many times. The high voltage winding normally goes nearest the core, and it is possible to alter the number of secondary turns on some transformers to change the secondary voltage.

The primary winding, the one that is first powered and magnetizes the core, needs sufficient turns to make its inductance appropriate to the applied voltage, so that the standby current will be minimal. The cross section area of the core is also a major part of the design equation, since that is what carries the power in the intermediate magnetic state. Transformers up to 50 watts for supplying small electronic modules often have about four turns per volt on their windings. For example, a step-down transformer from 120 to 20 volts might have 480 primary turns, and 80 secondary turns.

In practice, small transformers have slightly more turns on their secondary windings to make up for core inefficiency, a factor that is balanced against cost, because a certain voltage drop is anticipated when the transformer is operated at its nominal rating of, say, 0.5 amps in the secondary circuit, and the off-load voltage can be up to fifteen percent higher. This is not the problem it might appear, because solid state regulators economically deal with the difference.

The number of turns per volt is the same for primary and secondary because:

THE VOLTAGE RATIO OF A TRANSFORMER IS DIRECTLY PROPORTIONAL TO THE TURNS RATIO.

The current ratio is inverse to the voltage ratio, in order to maintain the same watts transfer from primary to secondary circuits, therefore:

THE CURRENT RATIO OF A TRANSFORMER IS INVERSELY PROPORTIONAL TO THE TURNS RATIO.

For example, if the load on a 120 to 20 volt step-down transformer secondary passes two amps, then the number of watts transferred is 20 × 2, or 40 watts. Ignoring losses and efficiency for the purpose of demonstrating the principle, this means that the primary current will have to be 40 watts divided by 120 volts, or a third of an amp. The product of primary volts and amps equals the product of secondary volts and amps; 40 watts of electricity in both primary and secondary; 40 watts of magnetic energy in the core. The equation between primary and secondary is always valid; if the secondary current falls to half its value, then so does the primary current. The magnetic core flux is a charged energy reserve which will supply what the secondary circuit load requires, and which reflects a similar demand on the primary circuit.

The calculations are then qualified by the efficiency of the transformer, which is probably between 75 and 95 percent.

Impedance is also reflected through transformers. Just as the turns ratio determines the voltage or current ratio of a transformer, the square of the turns ratio governs its impedance ratio. Transformers correctly match circuits of different impedance, but whereas the voltage ratio of a transformer is directly proportional to the turns ratio:

THE IMPEDANCE RATIO OF A TRANSFORMER IS PROPORTIONAL TO THE SQUARE OF THE TURNS RATIO.

AUTO-TRANSFORMERS

The majority of transformers fully isolate the two circuits they serve, but there exists a sub-class of transformer that employs a single tapped winding, or two windings of suitable conductor size for the relative currents involved, joined end to end. Where economy, efficiency, compact size, or variable output is required, the auto-transformer is used for both power and audio. In power mains applications, an insulation barrier is not required as there are no separate windings, but the transformer does not provide the safety of isolation. Mains voltage therefore is present at the transformer output relative to ground, but there are many applications where this is acceptable.

Power auto-transformers are generally used for voltage ratios up to two to one, as extra low voltage uses are usually for equipment that has to be fully isolated. In audio, line to speaker impedance transformers are commonly auto-wound transformers, or "tapped chokes." Variable power transformers, tapped with a movable carbon brush, are examples of auto-transformers. Because the variable transformer must have an accessible winding deployed in either a straight line or a circle so that the output contact can sample it along its length, it is wound on a ring-core, the most efficient type, enabling it to have the minimum number of turns. The output brush is shaped to minimize adjacent turn shorting, which would cause heating and loss of efficiency. As shown in Fig. 11.4, it has an intermediate tap so the transformer can be connected either for a maximum output which is the same as its input, or for an available overvoltage of about ten percent.

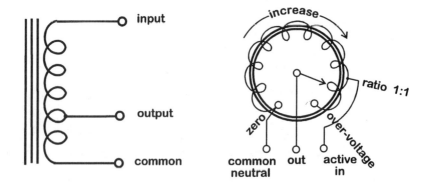

Fig. 11.4 Auto-transformer (left), and variable transformer.

MAKING AUTO-TRANSFORMERS TO ORDER

On odd occasions the need will arise for an adjustment transformer to marginally change a supply voltage. Although a 120 to 105 volt transformer would be a special order if purchased from a supplier in the normal way, the auto-transformer principle will provide the means, using standard components (Fig. 11.5). 105 volts is 120 minus 15, so a 120 to 15 volt step-down transformer is selected with both windings rated not lower than the current you wish to draw at the load, and the primary and secondary are connected in series. The output will be either 135 or 105 volts, depending on the phasing of the two windings, so a voltage check has to be made before applying the load or finalizing construction. If the 15 volt winding used is the secondary of a normal step-down transformer, then its wire gauge will be heavier than the primary and able to carry the current involved.

Bear in mind that this project deals with power mains that are not isolated by the transformer. Any winding not double insulated from the series secondary can not be used for any other purpose. Mount the transformer in a grounded metal box, or a plastic junction box of suitable strength, and provide internal terminal strips and cable restraints. Make sure any windings or terminals in the mains circuit have adequate insulation from the core and the mounting brackets, and preserve the principle of double insulation where mounting screws are accessible, unless they are grounded.

If the load equipment has a chassis or case that must be grounded, then ground continuity has to be reliably carried through from input cable to output socket. Ventilation is needed, and if there is any doubt about the current rating or enclosed temperature rise, give it a soak test; run it under load for a couple of hours, and observe it.

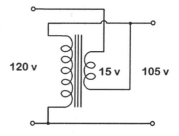

Fig. 11.5 Using a low voltage winding to make a low ratio auto-transformer.

TRANSFORMER FREQUENCY RESPONSE

All transformers need to be efficient at the frequency of their use, including power transformers, which must have an adequate core to prevent magnetic saturation. Mains frequencies are different around the world, and transformers and electric motors have a design frequency and temperature rise. If either are used at a frequency too different from that that specified, they will run hot and risk failure. More iron in the core is required for lower frequencies, and power transformers are designed to be efficient around the lowest mains frequency they are to use. Transformers for switch-mode power supplies use higher frequencies, typically 300 Hz, and this means that other core materials which are not efficient at mains frequencies can be used to build a much lighter transformer.

Audio transformers for the range 20 Hz to 20 kHz have cores that work adequately at the low frequencies via the core only, but need special techniques like winding inter-leaving (Fig. 11.6), which increases the direct coupling between windings, to transfer the higher frequencies. Audio transformers are manufactured for their particular application. A microphone transformer, for example, can not be used at line level because the core saturates and distortion results.

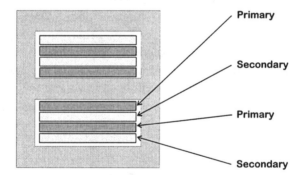

Fig. 11.6 Showing the method of winding inter-leaving to increase the transformer coupling at high frequencies.

THE ELECTRICITY SUPPLY MAINS

Power distribution throughout the world basically operates the same way; only the voltages and frequencies vary. Three phase mains are normal, although multi-phase and two-phase systems are used industrially. General distribution is via high tension mains to keep current and cable weights relatively low. After a local transformer steps down the high voltage mains to the three phase consumer voltage, supply lines have the neutral **Star Point** solidly grounded, although there can be a certain voltage from neutral to ground, making it important that neutral lines should not be grounded inside equipment or sub-mains wiring. Three phase waveforms are shown in Fig. 1.5c. One effect of the neutral ground is to guarantee that the three phase voltage, which is higher than the single phase voltage, will never appear anywhere relative to ground. It is therefore safe to supply different equipment within a rack from different phases of the supply, but better to avoid it if it is practical, because it complicates noise loops and the type of noise they produce. In individual audio modules, most of the concern with power mains is converting the single phase supply to a suitable voltage, in most cases lower than the mains voltage, plus conversion of AC to DC; a process called rectification.

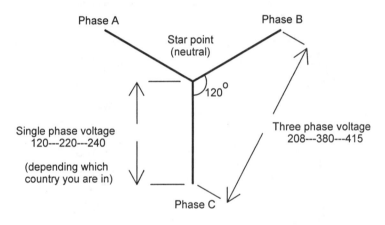

Fig 11.7 Relationship of three phase and single phase electricity supply mains.

RECTIFIERS AND POWER SUPPLIES

A diode passes current in one direction only. A circuit connected to an AC source through a single diode will have DC flowing though it with a high ripple content, because a single rectifier diode only passes half cycles. There are several ways to make both positive and negative half cycles flow in one direction from an AC source (Fig. 11.8), but the most obvious is the bridge rectifier, which has four diodes connected in a circle.

The other method, which ideally suited vacuum tube rectifiers, is a circuit that uses only two diodes, but requires a center tapped transformer secondary winding. The duty cycle of the winding is halved, as each side only carries half a current cycle, but there is the extra cost of a special transformer. Another two diode method is the **diode pump circuit**. This

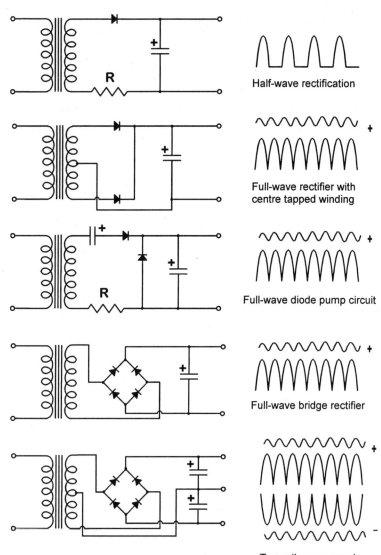

Half-wave rectification

Full-wave rectifier with centre tapped winding

Full-wave diode pump circuit

Full-wave bridge rectifier

Two rail power supply

Fig. 11.8 Power supply configurations, showing un-filtered DC waveform, and DC after filtering. Further ripple reduction is obtained by decoupling or passing through a regulator.

method produces full wave rectification at twice the transformer secondary AC voltage, from a single secondary winding, but regulation is poor and the output voltage is more load-dependent.

Many rectifiers use a bridge with four encapsulated diodes in one assembly with a common heat sink. Bridge rectifiers are often drawn in a diamond shape for identification. All sine wave rectifiers have ripple in the output, requiring smoothing before they will do anything except charge batteries. Even DC motors need a smooth power supply if they are to work efficiently, and above all, quietly. A high value capacitor placed across the output of a bridge rectifier acts in two ways. First, it's a filter, averaging the peaks and dips of the rectifier output. Secondly, it's a peak sample and hold device, bringing the rectifier output up from the RMS value of the original AC, to peak value, or RMS × 1.414, the inverse of 0.707, the relationship of RMS to peak waveform voltage described on pages 36 and 37, in regard to amplifier power output. The relevant diagram is repeated in Fig. 11.9 to assist explanation.

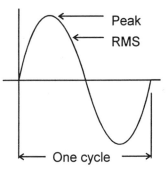

Example: A 12 volt AC transformer output will produce 12 × 1.414 volts of filtered DC if the capacitor immediately follows the full wave rectifier.

Fig. 11.9 **Elements of a Sine Wave.**

Large value capacitors charge and discharge fast. A series resistor is often used to limit the waveform peak charge current, which can be higher than the load current, and is a good idea for single diode circuits. A five ohm resistor can be used in series with small transformers, and temperature dependent resistors are used to limit the high in-rush current of toroidal transformers, which have a higher **Q** (resistance/reactance ratio) than other types and establish their field flux faster. Depending on the value of the capacitor, current can be drawn from the power supply up to the limit of its ability to supply the load, but if the load drains the capacitor faster than the ripple waveform peaks can replenish it, the voltage will fall towards the RMS value of the AC voltage, and ripple content will rise. In practice, power supply capacitors are chosen to be larger than the requirement.

Most amplifiers operate with plenty of available current, but no regulation, to supply the output stage, but either a regulator or a de-coupling circuit supplies earlier voltage amplifier stages (Fig. 11.10). Decoupling

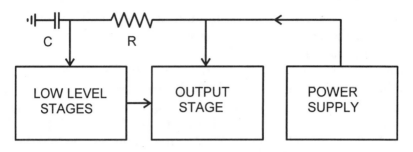

Fig. 11.10 Demonstrating direct power to the high level stage, and decoupled supply with voltage adjustment to more sensitive amplifying stages.

drops the voltage to the requirement of the circuit, and buffers the supply to early stages so that output stage signal can't feed back along the power supply rails and cause instability. The decoupling network is a low pass filter, and it further reduces ripple in the supply for the benefit of early stages where signal level is lower. The value of C is usually large, around 4700 microfarads, and R depends on the voltage drop required; it can be found by trial, or by calculation if the current consumption of the stage is known or estimated, using the Ohm's Law formula (Chapter 1 and Appendix A).

WHENEVER RESISTANCE IS CALCULATED, THE FACTOR INVOLVED REPRESENTS VOLTAGE DROP ACROSS THE RESISTOR, NOT THE VOLTAGE THAT WILL BE APPLIED TO THE FOLLOWING CIRCUIT.

REGULATORS

Regulators perform both the voltage drop and interference buffering functions of the decoupling circuit in a better way, by maintaining constant voltage or current regardless of load. Fig 11.11, shows a typical regulated power supply. The 2.7 volts held by the regulators is minimum. If the mains voltage falls by more than 15 percent, the regulators will be passing un-regulated voltage, containing ripple. The most a regulator should have to hold is about 30 percent over-voltage. Regulators can be cascaded to accommodate higher voltages.

In Fig. 11.11 there are two sets of filter capacitors. The first, following the rectifier, are large capacity for reducing ripple, usually 4700 mfd or higher. The second are 0.1 mfd ceramic disc capacitors which bypass RF

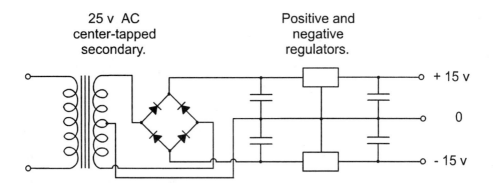

25 v AC
center-tapped
secondary.

Positive and
negative
regulators.

+ 15 v

0

- 15 v

Example:
   25 v AC delivers 25 x 1.414 v DC, = 35.4 v
   Centre-tapped 35.4 v delivers 17.7 vDC to
   regulators. Regulators hold 2.7 volts.

Fig. 11.11 A three-rail regulated power supply.

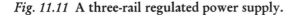

better than electrolytics, fitted physically as close as possible to the low level circuit components. They are there to shunt any RF to ground in the interests of stability and to attenuate high frequency regulator noise (page 228).

There is usually no requirement to regulate the supply voltage to the power stages of an audio amplifier, since an adequate signal to noise ratio is obtainable without it and voltage regulators for high current would be an unnecessary sophistication.

VOLTAGE AND CURRENT REGULATION

A three terminal IC regulator can also be connected to regulate current. Typical uses are for constant current Nicad battery chargers, constant voltage chargers for lead-acid batteries and over-current protection for servo amplifiers and the DC motors they are driving.

Better low speed performance is obtained from small DC motors if they are supplied a constant voltage. They accelerate to full speed immediately and are less load dependent than motors with an unregulated supply.

Current regulators permit normal operation of an amplifier up to the limit of the set current, so maximum acceleration of a servo-motor is available, but as load impedance falls sharply as the motor approaches standstill, current is held by the regulator to a maximum of, say, 5 amps, so that a 100 watt 24 volt motor will not burn out under low-speed/high-torque conditions.

ZENER DIODES

Integrated circuit regulators evolved from simple circuits using zener diodes. The diode acts as a selective shunt across the voltage controlled circuit, and the voltage dropping resistor connects it to a supply of higher voltage often dissipates a lot of heat in the process. Zener failure could result in full voltage being applied to the following circuit, with damaging results. It is possible to extend the current capability of such a circuit by having the zener control a power transistor. Later, integrated circuit regulators came on the scene which were efficient but fragile; a momentary overload would often result in instant failure. Modern IC regulators have short and long term overload protection and high temperature shut-down and recovery, but it is wise to operate them with a fair margin, otherwise they may start shutting down prematurely when they have aged.

While it is possible to make a regulator with discreet components, in most cases it is much better to use the large range of IC package regulators which come in a range of output currents and voltages in both positive and negative configurations. The larger regulators in a TO-3 case, for example, will pass several amps. The current capability of regulators can be extended by adding power transistors, but there is some loss of precision.

REGULATOR NOISE

The exception to using IC regulators, when it can pay to make a regulator from discreet components is the occasion when a very low noise power supply is required, as for some preamplifiers. Zener diodes, the voltage sensing elements used both in discreet and in IC regulators, are **backward operating devices**; that is, they oppose the flow of current below the zener voltage, when they start conducting to shunt over-voltage to the zero volts rail. The break-over at the zener point is inherently noisy.

Ordinary LEDs, light emitting diodes, are **forward acting devices**, and have a constant voltage drop of about 1.7 volts and are therefore relatively noise free. When used as the voltage sensing element in a regulator circuit, it is no trouble to to devise a network which translates their constant voltage into a higher one, which then controls a power transistor.

SWITCH-MODE POWER SUPPLIES

Mains frequency transformers are heavy and expensive, but there exists another class of power supply which switches a pulse width controlled current through a smaller transformer at higher frequencies. The transformer often has a ferrite core, efficient at the higher frequencies used, such as 300 Hz. The outcome is a very much lighter power supply with better controlled power-up characteristics.

THREE DOLLAR TESTER

The use of a light bulb in series with the mains, for safe testing of equipment in doubtful condition, has already been discussed in Chapter 3 in relation to amplifier testing, and is worth repeating in this chapter on power supplies.

However, there are exceptions. A series light bulb may be no safeguard when testing switch-mode amplifiers, or in any situation where large power supply capacitors may be discharged rapidly by system faults. Charged capacitors can store an enormous amount of energy, and should the inhibit circuits fail, allowing both positive and negative power rails to acquire a path to the zero volts rail at the same time, the capacitor discharge can cause components to explode with great violence, and it is preferable to give these service jobs to a specialist. Service technicians familiar with hazardous equipment make sure the covers are on before powering up.

The power mains has good regulation; if a short circuit occurs in a power supply or its following equipment, the mains voltage will remain at its nominal value, and cause high current to flow though the transformer and following circuit. For this reason, fuses protect the system. Very often, a situation arises where it is difficult to diagnose the cause of a blown fuse, and the handiest thing anyone can own is a dummy fuse, consisting of a blown fuse cartridge wired to a 100 watt light bulb.

When the power is turned on with the series lamp in place, full supply voltage will be available in the quiescent or no signal state, and the equipment will operate at low level, enabling diagnosis to proceed, because the series lamp will limit current to less than one amp.

If everything is OK at switch-on, the lamp will flash briefly as the transformer in-rush current establishes the core flux and charges the filter capacitors, settling to a dull glow, or at worst, half brightness. But if anything is seriously wrong, the lamp will light to full brightness, indicating a short circuit, and limiting current to a non-destructive level. The saving in replacement fuses and frustration is worth having.

A more adaptable version of the dummy fuse is a light bulb in series with a power plug and socket, with a switch to by-pass the lamp, allowing unlimited current to flow if everything checks out. This is a handy and necessary part of an amplifier service kit, as it provides a safe and quick check of an amplifier that may have been dropped or suffered a serious fault. The amplifier will work with the lamp in series, but only at low level. So there's no need to be afraid of most pieces of equipment that blow fuses violently, or have a history of going bang in your face, with the exception of the special conditions mentioned previously.

A series lamp not only provides safety, but also valuable diagnostic information when checking out equipment that blows fuses at specific parts of its operating cycle. A typical example is a jammed micro-switch or relay, which may short-circuit the supply for half a second or less. In this type of situation, the momentary flash to full brightness indicates the exact timing of the fault, speeding diagnosis by pointing to the part of the circuit containing the error.

When dealing with misbehaving control systems far from home at all hours of the day and night, a technician needs all the help available; and it is just as welcome from a tool that costs almost nothing.

POWER SUPPLY FAULTS

The following components fail in power supplies:

The transformer primary is on the inside, so it gets less convection cooling. A step-down transformer primary also has thinner wire, so primary winding failure is more common in small transformers because they are intentionally fragile, and are designed to fail safe instead of starting a fire. Some transformers have a thermal fuse in the primary winding that can be bypassed in an emergency, as long as there is a back-up fuse.

Diodes can fail, especially the discreet types which have lower thermal inertia. Consider that they do a fantastic job; stressed positive and negative many times a second for several years. An open or short-circuited diode can be detected with the ohmmeter probes without removing it from circuit. It is mentioned in Chapter 2 that a cheap analog meter that draws a certain amount of current may be a better diode tester than one with higher impedance. Most digital meters have a special range for diode testing.

Electrolytic capacitors, which are used in power supplies due to their high capacity in a relatively small package, are a semi-liquid device, and after about ten years of faithful service, they can dry out and lose capacity, or short-circuit. Routine replacement of filter capacitors used to be recommended some years ago, but reliability has greatly improved, and it may not always be a good idea, particularly since they are an expensive item.

Wherever this symbol appears on any service diagram, power supply, or elsewhere, it means that a component has a safety function and must be replaced with one of the same type. Safety components are designed to fail first to prevent further damage or fire.

Safety standards around the world are not the same, so fuse ratings and use of safety components should be observed unless the equipment manufacturer permits variations. Note that fuse cartridges are rated in voltage and case type as well as current and blowing time; fully enclosed fuses should be replaced with the same type.

Valuable equipment is worth additional protection in the form of mains surge suppressors. Computers, VCRs and stereos; even refrigerators, respond well to suppressors worth only a few dollars when faced with indirect lightning strike and gross switching transients. It's good insurance.

BATTERIES AND CHARGERS

The two most common re-chargeable storage battery types used in portable electronic equipment are Lead-Acid, and Nickel-Cadmium, and they have individual and distinctly different charge and discharge characteristics.

LEAD-ACID BATTERIES

Lead-acid batteries, like car batteries, are capable of rapid charge and sustained high current discharge. To improve their suitability in portable equipment, they use an acid-gel electrolyte, which renders them maintenance free and non-spill. Ordinary liquid-electrolyte acid batteries out-gas in use, producing an explosive mixture of oxygen and hydrogen, and since these are the elements of water, the water content of the acid solution has to be replaced with de-mineralized or distilled water as it is lost. However, low-maintenance and acid-gel batteries last between two and five years without needing water replenishment, and gassing is minimal.

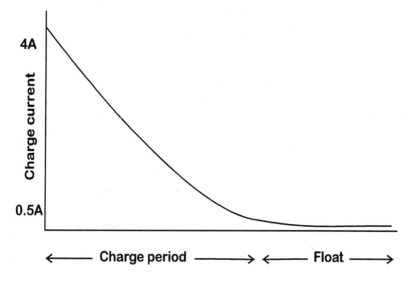

Fig. 11.12 Typical charge characteristic of a lead-acid battery.

Lead-acid batteries of all types require **constant voltage** charging, and prefer a charger regulated to a specified "float" voltage; 14.2 volts for a six cell 12 volt car type battery, and down to 13.4 volts for compact gel type 12 volt batteries. The acid battery is nominally two volts per cell.

The term "float voltage" (Fig. 11.12) refers to the point above nominal battery voltage where it is designed to "float" across the charging voltage, charging the battery, then maintaining its state of charge as current falls to a minimum. 12 volt car batteries float at 14.2 volts, whereas smaller acid-gel batteries in portable equipment float at around 13.4 volts, or 2.233 volts per cell.

On a constant voltage charger, the battery charges rapidly at first, and falls to half an amp or less as it reaches the float voltage (Fig. 11.12). Over a further period, charge current falls to minimum for the battery type and condition. In practice, a car battery on "trickle charge" passes about 250 milliamps to maintain a state of full charge. The measure of a lead-acid battery's state of charge is its terminal voltage under a small load, and the specific gravity, or density, of the electrolyte. The measure of the battery's condition is its ability to maintain its nominal voltage under maximum load for several seconds after a normal charge period.

Lead-acid batteries of all types can be left on charge for extended periods without damage provided the float voltage is not higher than specified and the water content is maintained. If the float voltage is below specification, the battery will not acquire a full charge. If the float voltage is higher than specified, the battery will suffer damage and its life will be shortened. Electronic battery chargers maintain a battery in better condition than simple chargers.

Acid-gel batteries are used in some UPS units (un-interruptible power supplies), where the charger and battery remain permanently on the power mains, and a sine wave inverter run from the battery supplies computers or any equipment that can not afford a momentary power failure. The battery of a 1 kilowatt UPS is about the size of the one in a cordless electric lawn mower, but the charger section of a UPS is much larger than its counterpart in the mower as it must be capable of delivering enough continuous current to supply the 1 kilowatt power drain on the battery circuit, rather than replenishing it over 24 hours after one hour of use.

PRECAUTIONS WITH ACID BATTERIES

Acid batteries suffer permanent damage if stored in a state of discharge, and should always be re-charged immediately after use. The deterioration process is called "Sulfating." Because they out-gas an explosive mixture of oxygen and hydrogen, liquid electrolyte lead-acid batteries must be protected from sparks. Chargers should be turned off when connecting and disconnecting the charger leads, and the battery should be well ventilated. An all too frequent occurrence when jump-starting an engine is to detonate the gas mixture in the vacant space between the electrolyte and the top of the case. The result is a battery without a case, and acid liberally sprayed in all directions.

The sulfuric acid in a battery is fairly dilute, and will not seriously burn the skin, but when allowed to evaporate, the water content disappears, leaving concentrated sulfuric acid, with its attendant dangers. For this reason alone, acid spills should be cleaned up immediately. If there is a possibility of spill or explosion, safety goggles should be worn. Airlines will not carry goods or luggage containing liquid electrolyte acid batteries.

Lead is a cumulative toxin; hands should be washed after handling lead-acid batteries.

NICKEL - CADMIUM BATTERIES

Nickel-cadmium or Nicad batteries are quite different to lead-acid. They are lighter, and more suitable than acid batteries in small packages of one or two cells. At a fully charged 1.25 volts per cell, their terminal voltage is lower than dry chloride batteries 1.5 volts, but more capable of operating at higher discharge rates. Their voltage falls from 1.25 to 1.2 volts per cell during the usable discharge cycle.

They require a **constant current** charge of approximately one tenth the maximum rated discharge current. This means that the charger's output voltage should be 30 percent higher than battery nominal, but limited by a current regulator or a series resistor set at mid-charge current. The charge current of a simple charger varies during the charge, but not as much as a lead-acid battery.

Nicad batteries can be left in a discharged state without damage. They do not gas or leak except under adverse conditions. The electrolyte is a corrosive alkali, and they should not be opened. Cadmium is a cumulative toxin which displaces calcium in bones, causing fractures, so contact should be avoided.

An adverse characteristic of Nickel-cadmium batteries is their "minimum discharge memory," in that the discharge level at which they are then recharged becomes the maximum capacity of the battery. This can be reset by two or three full discharge and charge cycles, but a better method is to use a pulse charger, which prevents formation of the low discharge point memory, and also enables faster charging.

CHOICE OF BATTERY

Lead-acid batteries are ideal for high drain applications, or equipment calling for a large total amount of energy. For the same amount of power, they are cheaper than nickel-cadmium batteries.

Nicad batteries are ideal for relatively high drain, medium power storage, but are quite a lot heavier than dry batteries. Compared to dry batteries, they also tend to go flat unexpectedly, and are not so suitable for long term power, as they self discharge slowly over a period.

Depending on the availability of replacements, dry batteries are very suitable for medium drain equipment like portable tape recorders, as their **no-maintenance** shelf life of one year plus is superior to either lead-acid or nicad batteries, and their characteristics are reliable and predictable.

There are many other battery types, like nickel iron and mercury batteries, which are not in such universal use for moderate power applications as lead-acid and nickel cadmium primary batteries, and chloride-zinc secondary batteries.

Chapter 12.

Control Systems

BASIC CONTROL GROUPS

In many systems or audio installations there is a sub-system which is not involved with audio processing, but which accomplishes remote function switching and distributes power to external equipment.

The control system integrates control functions into a central command area, or to distributed command points, and provides automatic interlocks between some of the sub-sections. An example of this would be the relay (Fig. 12.1) that cuts the studio monitor speaker when a microphone channel is opened, or the automation that dims the lights and opens the curtains in a motion picture theater.

Controls fall into one of three groups:

1. Power Control; mains voltage to audio equipment, projectors, curtain motors, stage and house lights, and mains voltage relays. This involves directing mains power to remote locations for the purpose of turning equipment on and off, running and reversing electric motors, and sometimes operating relays or contactors with mains power, the last being sometimes necessary to integrate a sound or auditorium control system with existing equipment, like the lights and curtain motors.

2. Low voltage control; relays and small motor functions. This is to control remote power or function relays, and is preferable to power control where the equipment being controlled is all part of a new installation or can be fully integrated. Solid state switches perform similar functions to mechanical relays. Their merits are discussed later in the chapter. Small motors are also used on occasion to operate switches and gain controls; the advantage being that they can

be operated from several parallel control points, if necessary, although this type of control method is being phased out by electronic methods.

3. Digital controls; a fast data stream to various parts of the system via one or more conductors is interpreted by the various decoders into individual functions. This is a hard wired version of the familiar infra-red or radio remote control. Digital control is certainly the way to go, as many commands can be multiplexed in a single line, but is easy to set up only if it comes integrated into new equipment. Examples of serial digital interface are the link between a lighting control panel and a bank of stage light dimmers, and the data line from the time code reader on a tape to the display modules around a studio that address to every portion of the program in hours, minutes, and seconds.

Computer programs are used extensively for controlling audio visual programs, automated cinemas, and studio mixers, far surpassing dedicated relay systems in flexibility. A dedicated, or single purpose automation system like the one shown in Fig. 12.3 can be compared to a traveler who follows a road network and is able to make decisions to turn left or right, start and stop. A more advanced traveler has a computer based system with the ability to alter the program at will while it is running. This traveler can take off and fly.

RELAYS

Relays (Fig. 12.1) are electro-magnetically operated switches that perform remote switching and simple logic functions, and they interface control systems of different voltages and currents.

Relays come in different current capacities and contact configurations, one of the main features of choice being the relay's suitability for very simple logic functions, like timing and sequential operation, or its use as an interface between the control system and the power circuits.

Relay coils come in almost any convenient voltage, like 5, 12, 24, 120, and 240 volts, and are available for AC or a DC. Small DC relays are used for relay logic applications. The following circuits show the relays applied to various situations, and although there are many options, it is often convenient to use circuits that use all "make" contact pushbuttons to latch them and release them. This is partly because of the popularity of

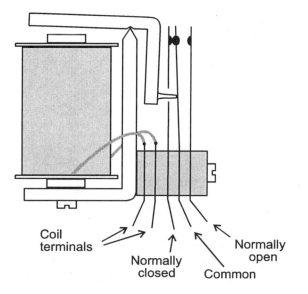

Coil terminals

Normally closed

Common

Normally open

Fig. 12.1 A basic relay, which may contain an average of four sets of change-over contacts like the one shown.

membrane switches (Fig. 14.16) which sometimes have printed circuit contacts, and can therefore only be made as "make," or "normally open" contacts.

Component values in the examples shown are for 12 volt, 320 ohm relays, but it should be noted that there are many types and brands of relay, and they will not all perform the same, as there are wide variations in their mechanical and electrical characteristics.

If there are lamps supplied by the holding contacts to indicate the state of the relay, and the circuit has been designed by trial and error without the lamps in circuit, the additional voltage drop across the holding resistor is often enough to cause the relay to drop out when the lamp loads are applied. It is therefore necessary to choose a holding resistor that allows for the drain of the lamp as well as the coil, or the whole system will have to be re-balanced. A 20 milliamp LED and its series resistor connected across the relay can draw as much current as the relay coil, unless a separate contact is used to supply it. Lamps have been omitted from the drawings in the interests of clarity.

Latching relays (Fig. 12.12a) are used for many on/off applications, as also are alternate action relays which maintain a changed state mechanically, like a push-on/push-off switch, except that relays can be operated

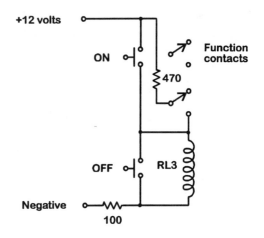

Fig. 12.2a The circuit on the right is the basic latching relay.

When the ON button is pushed, the energized relay closes and is held in that state by a smaller current supplied through the holding resistor.

A latching relay could be turned off by opening a *normally closed* contact anywhere in the coil holding circuit, but to enable the use of *normally open* pushbuttons, a low value resistor has been added so that the coil can be momentarily shorted to make the relay drop open safely, even if the ON button is pushed at the same time.

A disadvantage of using normally closed cancel buttons to interrupt the coil circuit is that a poor contact in the button can cause the relay to drop open prematurely.

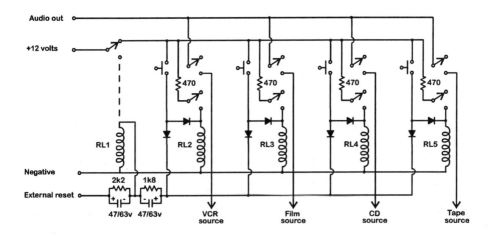

Fig. 12.2b A bank of self cancelling relays is shown in a circuit based on that of Fig. 12.2a.

The diodes isolate the on-buttons from each other, and the drop-latch relay RL1 is timed to blink briefly to cancel previous latches.

Selecting a relay therefore drops out any others currently selected, and only one latches at a time. The relays can also be reset by a momentary command from external equipment. Diodes isolate the relay circuits from each other.

by remote control, or by logic circuits. A self-cancelling relay (Fig. 12.12b) array is useful for selecting different functions like audio sources or motor-driven curtain and screen masking positions, for example. For this type of control system, a momentary "blink" relay (RL1) is used, which drops out any other latched relay before the pushbutton is released.

If electronic timer circuits are considered for a job, first read "Relays versus Electronic Controls" on page 245, because the last time I built electronic timers for a control system, the transistors failed when the power was applied to the mains circuits being switched, and I had to go back to the old method of using capacitors to delay the relays in order to meet the job deadline.

There are commercial timers that would have been suitable, but the job had been quoted and I was cornered, since they were more expensive. I am not going to say that you should not make your own solid state timers, but if mains voltages or high currents are involved, then it will be necessary to glitch-proof the timer and control circuits so that they will be immune to voltage transients.

RELAY TIMER CIRCUITS

The pulse circuit used to blink the drop-latch relay shown on the previous page is just one of several simple relay timers. The pulse circuit works as follows:

An electrolytic capacitor of fairly high value is chosen to store a charge at the voltage of the supply. On command, current flows freely, impeded only by the coil resistance, until the capacitor is full. That period of current flow pulls in the drop-latch relay. When the capacitor is full, the flow stops, so the relay drops out again, but it has done its job; the drop-latch relay has interrupted the supply to the relays long enough to cancel any that were latched.

The drop-latch event is over before the operator has released the button that initiated it, thereby selecting another relay, which latches. So the system works as a timed sequence to select, for example, a different audio source, while de-selecting the first. The timing can be adjusted to **make-before-break**, or **break-before-make**. The purpose of the resistor across the timer capacitor is to bleed away the charge in less than a second to ready it for another drop-latch cycle.

Fig. 12.3 shows other timer circuits. In the first, a high value capacitor across a relay short circuits the coil, delaying closure until it has filled. The capacitors across RL2 and RL3 fill slowly through the series 22 ohm resistors. RL2's closure is delayed for 6 seconds; the time it takes to fill the 9400 microfarad capacitor with a 320 ohm coil across it, through a 22 ohm resistor. RL3 is delayed for 3 seconds. The component values of these networks could be found by calculation, but the trial method is quick and accurate.

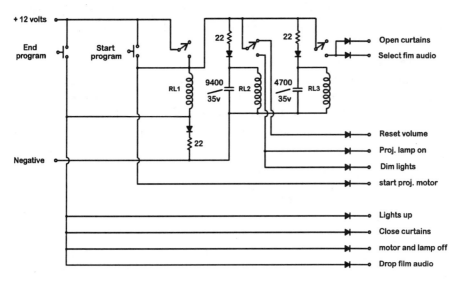

Fig. 12.3 Automation circuit using delays to close relays 1, 2, and 3, at zero, six, and three seconds respectively.

The function of this circuit is automated start-up and shut-down of a film projector, with sequentially timed house lights, curtain motors, projection lamp, audio select, and a center-reset on the motor-driven local/remote four channel volume control. When the system was built, it was a simple matter to run parallel controls into the theaterette as well, which really excited the owner, who had requested it.

MOTOR REVERSING

Where a control system governs power equipment, there is opportunity to include built-in overload protection and safety switches. One of the essentials in a motor reversing relay pair is the need to prevent both relays

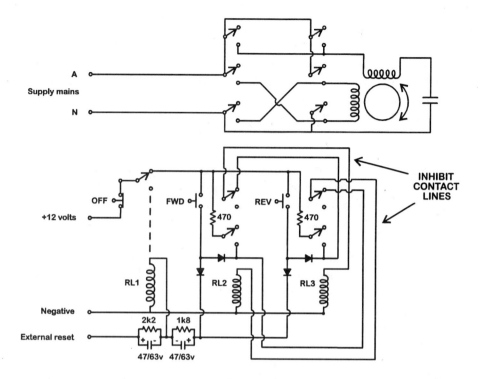

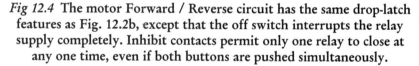

Fig 12.4 The motor Forward / Reverse circuit has the same drop-latch features as Fig. 12.2b, except that the off switch interrupts the relay supply completely. Inhibit contacts permit only one relay to close at any one time, even if both buttons are pushed simultaneously.

closing at the same time, as this would short circuit the power mains. Fuses will protect the system from major damage, but it is still not a good idea to risk extreme fuse overload. Even fast-blow fuses take a finite time to open the circuit, so they should be regarded as the line of last defence rather than an all seasons automatic circuit breaker.

The circuit in Fig. 12.4 shows forward and reverse controls, and the essential inhibit contacts that make sure a second relay can not initiate if one has already been activated.

SOLID STATE CONTROL DEVICES

Relays, diode pathways, and timing capacitors are the forerunners of diode and transistor logic, and just as these gave way to **TTL (Transistor-**

Transistor Logic), so TTL has been essentially replaced by microprocessors, and complex CMOS integrated circuits.

TTL is now used mainly for interfacing, since it has a relatively high current drain, and the early digital control systems using it often had a heat problem. CMOS runs cool, and is very much faster. Just as nobody today would make a control system using discreet transistors, so too, the trend is to program a microprocessor instead of designing an IC logic system.

The subject treatment will stop short of detailing the use of microprocessors and computers, because the diversity of the technology separates it from the study of basic principles. Application details of the large existing range of such devices are available from manufacturers and distributors, as well as very comprehensive manuals written by specialists in the field of electronic control. In support of this control concept, entire solid state interface assemblies are on the market, such as solid state relays, which make it very simple to switch the power mains from a 5 volt control circuit capable of delivering very little current.

Not very long ago, before transistors and computers came into universal use, the world's automatic telephone systems were service-intensive networks of single and bi-motional magnetic relays. If all the relay service staff had stopped work, the system would have started to fail immediately, most likely collapsing in a few weeks, whereas the immensely more sophisticated electronic modules on the whole do not go wrong, being subject to obsolescence rather than replacement due to failure.

This is not to say that there are not other sensitive areas in the system, but it serves to illustrate the greater reliability of electronic control which, once set up and de-glitched, has a very long and trouble-free life. Automated cinemas demonstrate the state of the art with computer control, and leave the operator only the task of initially loading the film on continuous loop machines and supervising their correct operation, as the computer starts each session at the correct time of day without an external command.

Opto-couplers are often used to isolate automation computers from power equipment, answering the requirement of all solid state control systems for the highest degree of immunity to voltage transients. Low current control systems like lighting dimmers are supplied already equipped with pulse sensing to dim the house lights up or down, and audio switching is accomplished with analog switches within stereo sound processors. However, it is most often a relay that interfaces a controller to a motor or high current device.

State of the art control systems technology requires that equipment is made to the highest reliability standards. One-off systems do not provide that assurance, and invite the decision to use simple, proven relay technology, or to opt for the high cost of a fully engineered and developed system with all the safeguards and no shortcuts.

RELAYS VERSUS ELECTRONIC CONTROL

There are many instances where a particular function can be performed with relays, and also with transistors or solid state logic devices. The essential questions influencing the choice are:

1. Do I have sufficient understanding and skill to construct and debug the solid state version?
2. How many do I intend to make: one, fifty, a thousand?

The first question brings up the comparison of the systems' relative immunity to voltage surges. Transistors and all solid state devices are susceptible to damage by transient over-voltage conditions. Just switching off a light can produce a momentary pulse several times the supply voltage. Back-EMF was discussed in Chapter 1 under Inductors, where one reason for the effect was given.

Commercially designed control systems have safeguards built in, like capacitors, inductors, and varistors, to bypass and block transients before they cause damage. Circuits also are designed for self-immunity.

The second question on the intended quantity queries the amount of time that should be spent reliably glitch-proofing the prototype, relative to the procedure's cost per unit.

It is quite true that as technology advances, so more users incline towards solid state techniques, and on the opposite side of the coin are those with a less high-tech. approach who have more success with relays, because they are simpler, they can see what's happening, and the devices are familiar.

Solid state is probably the way to go if the production run is more than five units. Once the design is established, it is far cheaper and more reliable than mechanical relay systems; and it is a lot more compact. Here are three case histories to demonstrate associated equipment failures:

1. A studio recorder does not record, even though the red light on the

panel says it is. With studio and performer time running at many dollars per hour, this is a memorable experience. It is found that an enclosed and factory sealed mechanical relay has failed to close completely because an iron filing has jammed the armature. The contaminant has obviously been in there since the relay was made.

2. A custom made solid state sound source selector in a cinema occasionally drops out for no apparent reason, sending operators running to restore the sound while the audience sits in silence. The cause is never identified, and is assumed to be one of the transients that produce a continuous background of hash in the power mains. It could be an air conditioner compressor switching in and out, but happens at random about twice a week. Perhaps it comes from outside the building. The source selector is rebuilt using miniature 5 volt relays and the problem does not occur again.

3. A mechanical relay control system is built for the picture and sound change-over function between one 16 mm and two 35 mm projectors in a preview theatre. It works well until an involved electronics engineer sees it, and expresses dismay that a simple function has been performed with a box containing five relays. It is replaced with a solid-state version which works perfectly until the first 35 kilovolt projection lamp igniter is triggered, whereupon the control immediately fails, and has to be re-built with additional safeguards.

The engineer is quite correct. Before transistors were available, control systems used relay-logic, an easy to understand system that carried the stigma of being mechanical. Once sold state controls are set up, and reliable transient protection is included, they are many times more reliable. However, a one-off design using relays is often an acceptable alternative where the budget is limited and only one such system is required.

Chapter 13.

Solder and Assembly Methods

SOLDERING

Compared to the other ways of making permanent connections, crimp, wire-wrap, and automatic terminals, solder is by far the most common method of joining conductors to components, printed circuit boards, terminals, and connectors. Although soldering is fast and flexible to experienced users, it is not as simple as it first appears. It needs instruction and a few pointers to become really easy and reliable. Even the mention of soldering might seem irrelevant to experienced people, but it's a fact that many people can't solder for nuts, even if they have spent considerable time in various departments of the audio industry, and their work regularly falls apart or goes noisy at moments of maximum inconvenience.

Solder is not perfect, and equipment that will be inaccessible for service often uses advanced assembly methods; like wire spot welding, and surface mount techniques, which require special soldering jigs. But for cheap, fast work, and normal equipment accessibility, standard soldering techniques are excellent provided the rules are followed, the nature of metal alloys is understood, and a brief time of practice is supervised.

TOOLS

The first requirement is a good soldering tool; either a temperature controlled soldering station or a portable iron with interchangeable tips, which has temperature compensation suitable to the size of each tip. The temperature of the join is critical, and the iron temperature will be higher in order to achieve this in the necessarily short application time. The wattage of the iron is not the deciding factor, as it is merely a means to achieve and maintain a suitable tip temperature as the iron loses heat into the environment. The tool manufacturer's specification will be the best guide. Elaborate solder/de-solder systems and special jigs for IC's and surface-mount components deserve a mention, but fall into the specialist category.

Avoid hobby irons; they are not reliable and have no temperature control, varying from from too cold to over-hot. Uncontrolled irons take a long time to reach working temperature. If you have to use an iron that gets too hot, power it through a drill-speed controller, and mark the **start** and **run** positions on the scale, or mount a series diode and a bypass switch in the handle or in an accessory box, observing the usual safety precautions when working with the power mains.

Some types of instant-heat irons pass high momentary current, and the back-EMF when they switch off or spark at the tip joint can kill transistors (see Back-EMF, Chapter 1). Propane quick-heat irons are very convenient for field service and free of back-EMF trouble. They are fairly controllable, with practice, and particularly useful for their portability and cord-free status. Suitable irons in most sizes have plated tips to prevent tin erosion. They should not be filed. Be warned that copies of good brands can be a trap; they look the same except for the brand name, and even if they operate at the right temperature, their tip plating may not stay properly tinned. Using a poor iron is a nightmare for a busy technician, and there is only one course of action; replace it with a good one.

Good quality hollow core solder contains a non-corrosive flux and at least 60% tin. Buy solder from an electronics supplier in preference to the nearest convenience store. Read the label, and never use 50-50 solder or acid flux for any electronic application. Soldering flux is a chemical compound that converts surface oxides into volatile chlorides and other by-products. Sheet metal contractors use solder composed of 50% tin and 50% lead, with a zinc-acid flux, but electronic work is done with 60/40 tin-lead alloy or better, and resin based acid free fluxes. If the items to be joined are not already tinned, then they must be perfectly clean, surface roughened if necessary, so that they will immediately acquire a bonded coating of solder, otherwise they should be tinned first.

SOLDERING TECHNIQUES

Heat is a matter of quantity, whereas temperature is one of degree. Connectors and small components can be damaged by application of heat for too long a period, as there is nowhere for the heat to go. Heat flows in for as long as the iron is in contact, and temperatures rise beyond specification in plastic-bonded and semi-conducting materials, so joins in electronics have to be quick-soldered to ensure they reach working temperature without cooking the components. This means that a cool iron needing a long application time will do more damage than the right tip temperature and a short application.

When soldering small connectors, the pins migrate out of position as the plastic supporting them becomes soft with the heat. They should be heat-sinked with a matching connector, or the pins pushed into a spiral wound with a few turns of copper wire. Both methods also serve as a work-holder that can be secured to a piece of wood.

Soldering at too low a temperature or movement of a join while it is cooling will produce a crystallized join in which the component metals of the lead-tin alloy separate into elemental crystals. If that happens, the join will have poor electrical properties due to the boundary effect of dissimilar metals. Oxidation will eventually follow, introducing another component into the join.

Over-long heat application will result in all the flux boiling away, and the join will oxidize. Oxidized or **dry joints** due to excessive heat or spent flux, and **cold solder**, the opposite cause but the same effect, are recognized by the dull appearance of the join, indicating crystallization. A good join is shiny and smooth, without excess solder, but enough to fill all parts of the join. If a join looks dry, remove surplus solder with the iron or a **solder-sucker**, a piston type hand de-soldering tool, or **solder-wick**, a copper braid made for the purpose, and re-heat it, feeding in fresh flux and solder.

Connecting systems normally have either a hole through which the conductor passes, or a cup or "solder-pot" arrangement which enables the conductor to be secured in a "well" of pure solder, which keeps the bonding areas clean and away from oxygen, while contaminants float to the outside. If connecting to a post, wrap a heavy conductor one full turn, or a light conductor at least twice, around the post, before soldering. Don't rely on the solder to hold the join together, because it is a conductive bonding agent, not a weld.

After the iron is **tinned**, that is, given a bonded coat of solder to make it "wet" enough to provide a heat conductive path to the join, solder is applied to the join components; not to the iron (Fig. 13.1). Don't carry solder to the join on the iron, because solder that stays on the iron has all the flux boiled off it. In cases where heat transfer to the join is poor,

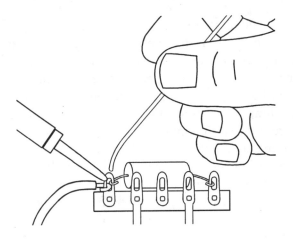

Fig. 13.1 Applying solder to the heated join.

application of solder to the junction between iron and joint is practical. Sufficient solder should flow into the join to leave the profile of conductors and terminal visible, unless there is a special reason for an over-filled solder join, as in a high current situation.

DON'T CARRY SOLDER TO THE JOIN ON THE IRON, BUT IF YOU HAVE TO, BE QUICK!

At the right temperature, solder flows like water, and should be given sufficient time, material, and heat, to fully permeate the join. Big joins can have solder fed in fast so that the join fills right up, but timely removal of the iron should give the liquid solder no chance to run out the bottom of the join.

SOLDER SHOULD FLOW LIKE WATER AS IT FILLS THE JOIN.

There is a class of miniature connector that has only close spaced short posts. Try wrapping two turns of conductor around a large needle so it will slip over the post, otherwise pre-tin the conductor and terminal, position them, wipe the iron clean on a wet pad, load the iron with solder and tap off the excess. Then quick-melt the tinned parts together, leaving the solder join slightly bigger than the two halves to make a strong bond. On assembly, make sure the cable restraint is taking all the strain off the connections.

Lead is insidiously toxic. It is a cumulative poison which causes all sorts

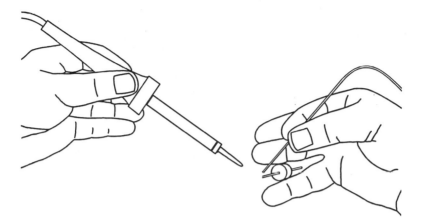

Fig. 13.2 Holding components and solder in one hand.

of hard to diagnose illness. The vapor from soldering should not be inhaled (a small fan blowing away from the job will extract it), but most contamination is from hand to mouth. I worked with a technician from interstate who used his teeth as a third hand, to hold the solder, of course. He was often ill, and suffered from chronic stomach pain. I was concerned enough to write to him explaining the risk as soon as I realized what was happening. Don't leave it accessible to children. If you've been using solder, wash your hands before eating. Choose to be free of this toxin, use all spare fingers as your "third hand" (Figs. 13.2 and 13.3).

There should be excess flux present after completing a solder join, and this can be removed with alcohol on a toothbrush if necessary, a particularly important point with printed circuit boards, as residual flux may contain contaminants, and attracts conductive dirt. Also check for loose whiskers of stranded wire and remove them. Multi-strand wires should be twisted before soldering.

Avoid soldering too close to a cable shield. A certain length of solder-free shield braid is necessary to isolate the heat and prevent short circuits inside the shield, as heat will travel along the conductor to melt the insulation. Some connectors are difficult to tin, and should be pre-tinned to avoid overheating the cable.

Laying the cut end of the wire flat along the printed tracks on a circuit board is recommended procedure where high current is involved, as in power supply boards, amplifier output stages, and speaker crossover networks. Similarly, high current circuit boards may benefit from building up the track thickness with solder to increase the conductor cross section.

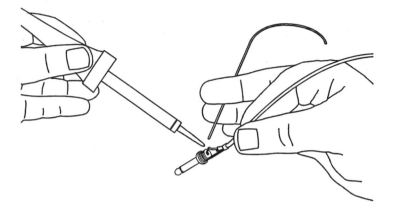

Fig. 13.3 Another way to manipulate three things at once.

When soldering resistors that will get hot in the course of their normal function, a suitable pigtail length should be allowed between the resistor and the join, or printed circuit board, so that the component will not melt the solder or char the board. Having seen a crossover network catch fire, I can categorically say that resistors that get too hot to touch should never be mounted in contact with a printed circuit board. Some manufacturers fit heat radiating aluminum stand-off supports about an inch and a quarter long on the leads of power resistors soldered to a printed circuit board.

Due to the risk of unseen short circuits between pins, the possibility of center conductors short circuiting to shields, and the ever present possibility of wiring errors, a routine of continuity and short circuit check is recommended after wiring every connector, other than on an assembly line. Use heat-shrink sleeving over closely spaced joins.

USING ADDITIONAL FLUX

The resin based flux in most cored solders is a balanced compound for the majority of assembly and service jobs, but there are components that defy attempts to flow solder on to their surfaces at a reasonable temperature. In these cases, use of a non-acid **paste flux** helps to initiate tinning at a lower temperature, which simply involves flowing a coat of solder on to a terminal's surface. Bare steel and some plated surfaces need paste flux to tin properly; the type used for capillary pipe fittings is suitable provided it is acid free. Tinning of steel is inhibited by the presence of some resin fluxes, and it's a simple matter to load the iron with solder, allow the resin to boil off, and then apply the dry solder to the flux coated job. Special fluxes are available for stainless steel.

Before tinning, the paste flux is applied to the job, not the iron, but a really dirty iron can be cleaned by plunging it at full working temperature into the flux pot and then wiping it on a wet pad. Residual paste flux residue should be removed from solder joins with alcohol, as it is sticky and attracts dust.

TROUBLESHOOTING, INSPECTION, AND RE-SOLDERING

Whether the solder joins on a PC board are new, or the subject of fault diagnosis in existing equipment, visual inspection is a necessary part of any operation involving solder. Many intermittent faults are cured permanently by re-soldering after inspecting every joint on the board with an eight times pocket magnifier. The whole point is that hollow or crystallized joins may work perfectly for two or three years, but oxidation will get them in the end.

What to look for is obvious once it is seen; cold soldering, dry joints, uncompleted soldering operation or parts of the join not fully tinned, and joints dirty or obstructed by flux residue. The magnifier is necessary because potentially troublesome flaws can be very small, and quite invisible to direct vision. Commercial PC boards, even from top makers, are made at speed on automatic assembly lines, and while they are fully tested before use, quality control is often imperfect, and undetected flaws do get through.

There's great satisfaction in fixing intermittent faults in a troubled audio or control installation that has defied service technicians for weeks, but which responds instantly to inspection and re-soldering as follows: Tighten every screw terminal, pull out every PC board and scan the solder side with a magnifier. Re-solder anything that looks suspect. Put it all back together, with special attention to making sure the PCB edge connectors are properly seated, and power it up. All installations respond well to routine service techniques. They have fewer breakdowns.

Don't trust solder until you have looked at it, but absolutely trust solder that you have just applied and inspected, because you know it will be reliable. The same comments apply to anything you solder; look at it; and if it is a shielded cable, and the insulated wire is not fully visible outside the shield for a quarter of an inch before the joint, check for continuity and short circuits. Looking at it also picks up any whiskers that stick out of untidy terminations. Wires need to be stripped to a minimum, twisted to keep all the strands in place, hooked around the lug to make a mechanical join first, and after soldering, inspected for faults and safety clearances.

Limited conductor clearance may be acceptable in miniature audio connectors, but it is essential to be safe if mains power or very big capacitors are involved. Power switches often don't have decent clearances to the equipment case, or the terminals may be accessible to wandering fingers. Solder joins get bigger and dimensions drift in equipment that has been extensively serviced, so even though the manufacturer may have considered it unnecessary, you can decide to make the equipment safe. Its a good idea, when in doubt, to add an insulating barrier of non-hygroscopic transformer paper.

WHEN NOT TO RE-SOLDER

Some devices are so heat sensitive that is inadvisable to touch them with a standard controlled-heat soldering iron. Generally they belong to the family of **LSI**, or **large scale integration** ICs, and **surface mount devices**. Integrated circuit chips and even resistors of this type are so fragile that application of heat in the standard way will damage them, and special irons and jigs are used in their assembly and service.

Many integrated circuit devices of the through-the-board variety are also heat sensitive, so the line has to be drawn between enthusiasm for re-soldering everything in sight, and a large dose of caution. Murphy's Law invariably makes a damaged component the most expensive and unavailable in the whole piece of equipment, so resist the temptation to re-solder a microprocessor or an unidentified IC unless you have another one handy. Some things should only be serviced in a specialised workshop.

CHOOSING AND FITTING CONNECTORS

Connectors are chosen for the following reasons, which are still valid when the installer has the option to choose:

1. Tradition, or Convention. Phono sockets on the back of home

stereo systems. These are very good provided a reliable type is chosen, and they don't have to work too hard; the all-metal phono plugs and sockets aren't always the best, sometimes the cheaper looking plastic ones are better. Professional users prefer microphone connectors for long service life.

2. Shielding and Mutual Isolation. Low level signals, RF and data signals that have to be contained, and power circuits that radiate low frequency magnetic fields as well as high frequency hash; all have their specific types of connector for the purpose of segregating their incompatible characteristics. It will be apparent that these circuits are also normally segregated by connector location and cable positioning.

3. High voltage or current. Power and speaker connectors are chosen for high current capability, reliability and safety. Its a fact that all connectors, including **Quick Connect** lugs and receptacles, perform consistently better if made of a material that has heat and wear resistant spring tension properties. Connectors with plated mild steel contacts that lose tension after a few uses should be avoided in favor of materials like brass and phosphor bronze. A loose contact will get hot if carrying any appreciable current. A hot contact spring becomes weak; tension decreases, and heating accelerates until the component fails. Check contact area too, when choosing connectors.

Tradition and convenience, however, sometimes win over suitability. The ordinary phone jack, developed more than a century ago for telephone switchboards, has become a music industry alternative standard for connecting speakers. Both plug and socket are fragile; made to poor dimensional and quality standards by non-brand manufacturers all over the world, and they have a microscopically small contact area. In their favor, however, they are cheap, easy to wire, and very quick in operation.

When fitting connectors, especially multi-circuit types, work out a reliable reminder to slip the cover on the cable first; if applicable. Its three times as difficult to wire a connector that has been un-soldered, so any routine for avoiding mistakes is legitimate. Saying pin numbers and cable colors aloud impresses them on the memory much better than just thinking about them.

On the subject of cable colors, it pays to stick to a standard sequence for multi-colored conductors. The resistor color code is standard throughout the industry. It basically follows the sequence of the white light spectrum as follows:

No.	Color	Abr.	No.	Color	Abr.
1	Brown	BN	6	Blue	BU
2	Red	RD	7	Violet	VI
3	Orange	OR	8	Grey	GY
4	Yellow	YW	9	White	WT
5	Green	GN	10 or 0	Black	BK

Soldering in a crowded row of pins is easy if the problem is thought through first. Choose the right iron tip, use small gauge solder, and above all, get plenty of light on the job. Soft light from a large source is best. Double check the order in which the pins are numbered before starting, and be clear on the pin order if it is not marked.

Draw the connector from the side at which you will be working (Fig. 13.4). Say **everything** about it: **Male Line Plug. Wiring Side.** Make sure the right gender is used. When people loosely say "Male. Female. Plug. Socket," they should be referring to the **functional circuit pins** of the connector, not the earth pin or the shroud of the connector, which may appear to be the other gender.

In matters of safety with power, high voltages, and circuits with high current capability, connector pins must be accident-proof; so the gender is chosen to make the circuit inaccessible to waving screwdrivers, falling paper clips, and wandering fingers, when there is nothing plugged in. People have been known to wire mains voltage to the wrong gender connector so that the exposed pins bite unsuspecting victims.

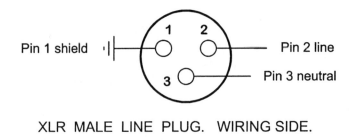

Pin 1 shield

Pin 2 line

Pin 3 neutral

XLR MALE LINE PLUG. WIRING SIDE.

Fig. 13.4 Draw the connector you are to solder. LABEL THE DIAGRAM SO THERE IS NO DOUBT WHICH *GENDER* AND WHICH *SIDE* OF THE CONNECTOR IS REFERRED TO. Pin numbers are not marked on all connectors.

SOLDERLESS AND CRIMP CONNECTORS

Solderless connectors come in many forms, but two that are frequently met in audio work are the XLR solderless connector and the crimp lug.

The solderless XLR works perfectly every time on suitable gauge conductors. Unlike the crimp, the wires are not stripped, but slide between contact blades that penetrate the insulation and grip the conductors under permanent tension. A small hand-operated assembly press fits the cap after laying the insulated wires in guides in the integral cable restraint. Once pressed on, the cap can be unscrewed and reassembled without damage. The system is so good that it calls for a word of praise for well designed connectors of every type, but advice to avoid connector brands that have weak cable restraints, or which come apart inside when the cable has a minor accident.

The illustration of a crimp-lug in Fig 13.5a demonstrates how the cable is compressed into an interference fit with the receptacle. The crimper recommended is a hand operated, double-toggle action plier with a variety of cable sizes represented, and a ratchet which ensures the crimp is completed before it will release. The temptation to apply another crimp on the other side of the receptacle should be resisted, as what's needed is a tight crimp, not a shattered one.

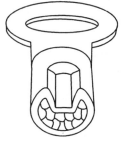

Fig. 13.5a Cutaway view illustrates how one type of crimp connector makes an interference fit with the cable strands.

There are many other crimp formats, but one that stands out as ideal for power connections is the **Quick Connect** system, shown in Fig. 13.5b, in which spring receptacles plug onto matching tabs. They come with or without insulation, and a variety of crimp formats. If a crimper is suspected of making an inadequate job, open type receptacles can be soldered as well, with a separate insulating sleeve for later application. Pull on the cable if necessary to verify each crimp.

One of the important assembly checks is to make sure the insulator is correctly fitted, so that the receptacle can be pushed right home on its tab, otherwise it may pop off at some later time. Use of a crimp connector system

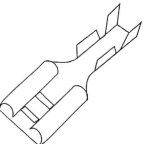

Fig. 13.5b **Quick connect receptacle, shown without the insulation.**

is dependent, as with all other termination systems, on care and inspection in use, and a quality control routine to guarantee consistent results.

WIRE WRAP

One of the fastest and most reliable types of connection, wire wrap (Fig. 13.6) solves many problems with low voltage wiring to circuit board wire-wrap pins and some specific components.

The contact area is large, and because there is no heat involved, oxidation is not encouraged, so current carrying ability is at least as high as the conductor capacity. The tinned copper wire is drawn around a square or rectangular post under sufficient tension to stretch the wire into the corners, producing an interference contact between the two metals.

Wire wrap is easy to do with the right size power tool for the purpose, or even a hand tool which consists simply of two holes in the end of a hardened steel shaft. The hole for the wire in the power version is usually a groove running in a sleeve, so that broken pieces of wire can be retrieved. Wire wrap is just as easily undone and reconnected, but about half the used wire should be cut off and the insulation stripped a short way back so that the start of the wrap is new wire. If this is not done, it will break sooner or later, having been weakened by the unwrapping from its tightly coiled state.

Wire wrap comes in different sizes, and is used for purposes as varied as speaker crossover wiring and computer circuit board interconnection. It is not the most compact connection system, but it is highly rated for convenience and ease of use.

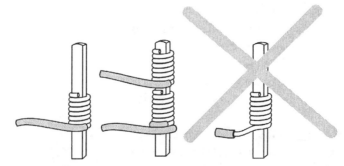

Fig. 13.6 Good and bad wire-wrap applications. The one on the right will be weak because the insulation has not been included in the wrap to ensure that the cable is anchored by the plastic covering.

Chapter 14.

Installation Planning

DIAGRAMS

An installation brief comes in many forms, from an interview to a written specification. It often does not come in a form that will enable the installer to directly proceed, without first translating or clarifying the job to be done. Even if the order comes with a written specification, then a set of drawings generally needs to be produced from it. This might seem a waste of time if drawings have not been asked for, but a lot of otherwise unavailable information about the hardware and the wiring can be derived from even a rough draft of the system diagram, although I concede that there are portable and straightforward installations that go together without a hitch or a single sheet of paper.

Block diagrams were introduced in Chapter 7. They generate the following essentials:

1. The drawing verifies the equipment that has to be supplied. Making a list from memory can lead to difficulties if an estimate is presented to the project manager with some costs missing.
2. Estimating the cable requirement and planning wiring layout similarly works better from a drawing. Fig. 14.2 shows the block diagram from which a **cable schedule** and a **patchbay diagram** (Fig. 14.1) have been prepared.
3. **Patchbays** are expected in many installations, even if they are not positively mentioned in the specification. An installation beyond a certain complexity is unworkable without a patchbay, and occasionally what appears to be described as a single **jack-field**, stated as serving microphones lines, line level jacks, and speaker lines, has to be interpreted as three separate patchbays. This is necessary because the different signal voltages and currents must be segregated, but also

because program flow is visualized along the length of the rack that carries the system hardware, usually from bottom to top, and there is a best location for parts of the system that have to be hands-on, like the central line-level patchbay.

4. Equipment racks and layout. Fig. 14.10 will show the method of allocating space to rack-mounted items, but it is the block diagram that provides the lead-up information.

Both cable estimates and patchbay layout are prepared directly from the drawing. Installation technicians don't always have well equipped drawing offices, and in any event they most likely would not have time to make a full set of professional drawings in the few days available to prepare for an installation. The drawing of Fig. 14.1 was made on A3 paper (16.53 X 11.69 inches), and reduced 70 percent on a photocopier for use and distribution. With the exception of the lettering, only a roll-ruler, a medium soft pencil, and a stencil of squares, circles, and triangles was used to make the drawing. If drawing a diagram like this one represents a challenge, not to worry, it didn't start like this, but as a rough layout in a technique fully described on pages 264 to 268. It is very easy by the method outlined.

If an ultra-neat copy is required for presentation, it can be traced and printed or simply photocopied. The drawing shown was in pencil, and was originally hand lettered, but when the project engineer insisted on printed lettering, I used typewritten paper cut-outs glued on to the drawing original as the fastest way to get it done in the limited time available.

A better way is to use stencils and a drawing pen, or better still, draw it on a computer using a CAD program, but without these things, many of us have to do the best we can with simple methods. The time frame for drawing preparation and analysis often makes this acceptable, but I do wish to impress on the reader that the larger jobs just won't come together unless a drawing of some sort is made. There's nothing quite so difficult as pulling more cables through a finished section of the building later because they were forgotten. The cable schedule and patchbay layout shown in Fig. 14.2 were worked out from the drawing. A cable schedule is many things to many people. It's an estimates list, to help when preparing quotations or ordering materials. It's also a working reminder list, especially in studio installations, where there are large numbers of lines of different function going short distances and then doubling back.

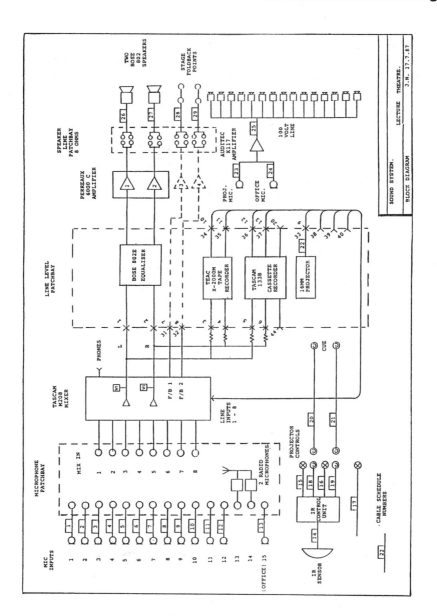

Fig. 14.1 The drawing from which the cable schedule and patchbay layout in Fig. 14.2 were prepared.

The cable schedule lists every minor line, even though it may start and end in the same room or console.

It is ultimately helpful to write complete cable schedules for those jobs to include all the cables in the rack, or around a mixer and the monitor lines, because a total schedule enables you to tick off each line as it is run. A typical example concerns the lines around mixers, with reference to Figs. 8.4 and 8.11, where the Dialog Group Return and the Tape-echo Generator reverb time control line simply connect one part of the mixer to another. Both functions could be performed with patchcords, but some dedicated or permanent setups are better hard wired. The amount of cable used in each case is minimal, just like cabling within a rack, so it is not going to affect the quote or the cable order list. But in terms of time wasted to cut open a cable loom, insert a forgotten line, and then tie it all up again, inclusion in the schedule of every single line, however short, has to be a good idea in this type of installation. I've worked with many clever people who can carry all the information in their heads, but I'm not one of them, and I avoid trouble by writing everything down.

The drawing will indicate which pairs of jacks should be normalled to achieve the ideal of a **dead jackfield**, meaning that it does not need a forest of patch-cords growing out of it to complete the "normal" circuit paths. The self-normalling switch-jack pair at the bottom of Fig. 14.3, also indicates that **tip**, **ring**, and **sleeve**, or **two-circuit jacks** are used in installations large enough to warrant a patchbay, so that line balance can be maintained as far as possible.

The Engineer's Specification will occasionally ask for no normalling. There will be a valid reason, to be sure, but when the finished system is handed over to the eventual owners, there is often a request for normals to be added. This may mean not only extra wiring, but jacks will also need replacement if they are not the normalling type, so jackfields should always be installed so that they are permanently accessible. The practical aspects of this are discussed in Chapter 10. Microphone and speaker patchbays are not normalled, and use different connectors. Note that in the system plan of Fig. 14.5, the choice of a speaker matrix type patchbay, requiring shorting plugs instead of patchcords, was done that way because it was specified, as also were Bantam jacks (page 270).

Normalling can be done vertically or horizontally within the one jack-strip. It is preferable to avoid normalling between two different strips, if only because it makes labeling difficult. Fig. 14.6 shows the detail of a designation strip for mounting on a separate panel above the vertically normalled jackfield.

CABLE SCHEDULE

CABLE NO.	TYPE	FUNCTION	ORIGIN	DESTINATION
1 – 4	TWIN SHIELDED	MIC. LINE	REAR THEATRE EAST	EQUIPMENT RACK
5 – 8	TWIN SHIELDED	MIC. LINE	REAR THEATRE SOUTH	EQUIPMENT RACK
9 – 12	TWIN SHIELDED ASC B1205 24/0.20	MIC. LINE	FOUR FLOOR POINTS THEATRE	EQUIPMENT RACK
13	TWIN SHIELDED	MIC. LINE	OFFICE	EQUIPMENT RACK
14	TWIN SHIELDED	IR SENSOR LINE	IR SENSOR	EQUIPMENT RACK
15 – 17	POWER 1 PHASE	REMOTE PROJECTOR POWER	PROJECTOR PATCH PANEL	EQUIPMENT RACK
18 – 21	6 COND. CONTROL	SLIDE PROJ CONTROL	PROJECTOR PATCH PANEL	EQUIPMENT RACK
22	TWIN SHIELDED	16MM AUDIO	PROJECTOR PATCH PANEL	EQUIPMENT RACK
23	TWIN SHIELDED	MIC. LINE	PROJECTOR ROOM	EQUIPMENT RACK
24	TWIN SHIELDED	MIC. LINE AUDIENCE RECALL	OFFICE	EQUIPMENT RACK
25	ASC A631 24/0.20 FIGURE 8 CODED	100 VOLT LINE	EQUIPMENT RACK	15 CEILING SPEAKERS
26 – 27	1 WHITE OLEX BAA 1 BLUE RO6 AA001 56/0.32	8 OHM LINE STAGE SPEAKERS	EQUIPMENT RACK	2 THEATRE SPEAKERS POINTS
28	1 WHITE OLEX BAA 1 BLUE RO6 AA001 56/0.32	FOLDBACK 1 8 OHM LINE	EQUIPMENT RACK	STAGE P.S.
29	1 WHITE OLEX BAA 1 BLUE RO6 AA001 56/0.32	FOLDBACK 2 8 OHM LINE	EQUIPMENT RACK	STAGE O.P.
30	FIGURE 8 24/0.20	BELL CONTROL LINE	OFFICE	EQUIPMENT RACK

Patchbay layout

Upper section (positions 1–24):

MIX OUT L	MIX OUT R	TAPE IN L	TAPE IN R	CASS IN L	CASS IN R	FB 1 OUT	FB 2 OUT	16MM OUT	TAPE OUT L	TAPE OUT R	CASS OUT L	CASS OUT R	MIX IN 5	MIX IN 6	MIX IN 8				CUE OUT	MULTIPLE A →			
1	2	3	4	5	6	7	8	9	10	11	12	13	14	15	16	17	18	19	20	21	22	23	24

Lower section (positions 25–48):

AMP IN L	AMP IN R	TAPE IN L	TAPE IN R	CASS IN L	CASS IN R				MIX IN 1	MIX IN 2	MIX IN 3	MIX IN 4	MIX IN 5	MIX IN 6	MIX IN 8				CASS CUE IN	MULTIPLE B →			
25	26	27	28	29	30	31	32	33	34	35	36	37	38	39	40	41	42	43	44	45	46	47	48

Fig. 14.2 Cable Schedule and Patchbay layout.

HOW FAR TO GO WITH DRAWING DETAIL

There are levels of drawing far more detailed than the block diagram technique, as the **Normalled jack pair circuit** at the bottom of Fig 14.3 shows, and these would include circuit diagrams, or schematics, and circuit board layouts, both of which show complete details of the equipment.

Block diagrams are for another purpose; the system overview, and to keep them simple, it is only necessary to show a single line for each cable, in fact, the less detail, the better. The Cable Schedule will specify the line type, but on the block diagram, a drawn line and its designation on the cable schedule indicates the minimum cable standard proposed in Chapter 3 for speaker lines, and in Chapter 10 for line level and microphone cables.

The set of drawings in Fig. 14.3 propose some alternate symbols for jacks, both open and normalled. The normalled jack detail at the bottom of the page is more at home on a schematic, or circuit diagram, and such detail should be avoided in a block diagram to keep it simple. However, detail drawings of connectors and jacks are often placed in a spare corner if an information key is needed.

Block diagrams were introduced in Chapter 7 as a means of getting to know complex equipment without having to view the entire circuit. Instruction was given to leave out non-essential lines, as they can be understood to be there. This still applies to a diagram prepared for cable scheduling, and later as a guide for operators, but in an installation drawing of this type it is obviously necessary to show a cable number on the drawing to indicate its presence.

DRAWING A SYSTEM BLOCK DIAGRAM

It is convenient to plan a letter size drawing, as it fits in a folder or operator manual and can be faxed and included in correspondence. However, except in the case of very simple drawings, it is best to start big, and then reduce it on a photocopier, greatly improving the appearance of the drawing. Convenient double size and letter sizes are:

ANSI B. 17" X 11"	*(431.8 X 279.4 mm)*	
ANSI A. 11 X 8.5"	*(279.4 X 215.9 mm)*	*(U.S. Letter size)*
A 3. 16.53" X 11.69"	*(419.86 X 296.93 mm)*	
A 4. 11.69" X 8.26"	*(296.93 X 209.8 mm)*	*(Metric letter size)*

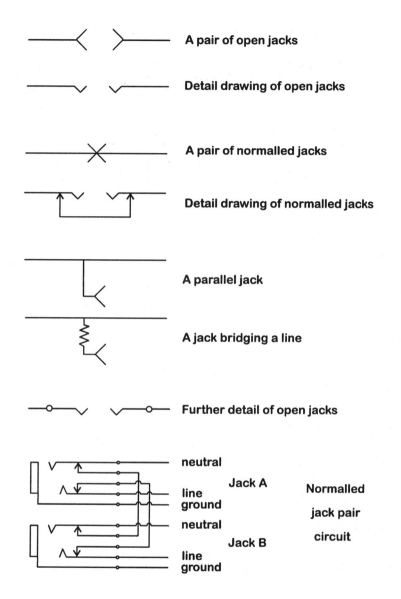

A pair of open jacks

Detail drawing of open jacks

A pair of normalled jacks

Detail drawing of normalled jacks

A parallel jack

A jack bridging a line

Further detail of open jacks

neutral

Jack A

line
ground

neutral

Jack B

line
ground

Normalled

jack pair

circuit

Fig. 14.3 Minimum and detail drawing styles of jacks and normals, for block and circuit diagrams.

A block diagram should not attempt to describe the physical layout of the installation area, or it will end up looking like a road-map of a few villages in a desert. If a plan view of the system is required, make it the subject of a separate drawing. Whether it is a rough sketch or a finished presentation drawing, it begins as a set of two or three rectangles representing the patchbays for the microphones, line level, and speakers, whichever are included, like the dotted line boxes in Fig 14.1. Make the middle rectangle big enough to contain other rectangles which represent the line-level equipment modules.

The drawing is built up around the patchbays, so that circuit events flow naturally from left to right across the page, and the amount of circuitry that does a U-turn and re-traces its path is minimized. So too are pieces of equipment that hover in corners of the drawing for no good reason except that there was room for them there. Laying out the equipment first is fatal; the diagram will tie itself in knots.

Having established the patchbays as boxes, all signal processing modules and sources with line level terminations at both ends are drawn as labelled rectangles inside the central line level patchbay. The lines and other equipment fall naturally into place. Don't worry at this stage if the layout is untidy, lines zigzag up and down, and items are not equally spaced, as there's a cleanup method which automatically resolves this problem in the final drawing.

An open jack socket is drawn as a "Y" on the border of the line level patchbay, and a pair of normalled jacks is therefore depicted as two interlocked Ys, which are drawn as a single "X." Opposite the top tail of each "Y" is the jack number, identifying the position it will be allocated on the jack-field. Labels can be applied to the jacks in almost any logical fashion, provided it is consistent throughout the drawing. A jack-field with more than two rows of jacks calls for a numbering system which gives the lateral position from 1 to 24 (or 1 to 48 for Bantam Jacks), vertical rows A to Z, and so on.

When the lines and main equipment modules are sketched in and nothing major is missing, write a set of numbers in the left and bottom margins for a vertical and horizontal grid indicating each line, the start and finish of each module, and the extra spaces between groups of different function within the drawing. The extra spaces are important, because if every item on the final drawing is exactly the same distance from the next, then the drawing will look like a wallpaper pattern, and it will not be easy to understand, no matter how neat it is. Function grouping makes a clear, readable drawing. Add another five spaces below the drawing for the title box and any required footnotes.

The available paper length and width inside a suitable margin is divided by the number of grid-lines. The grid is then redrawn, equally spaced, on a fresh piece of paper. Draw the grid in bold black pen, and use it as an

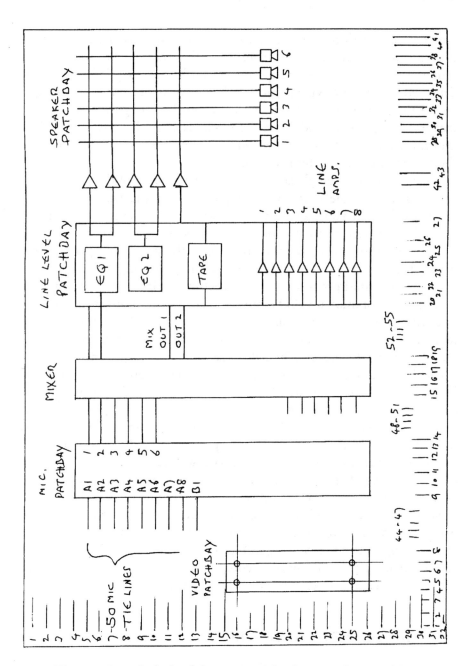

Fig. 14.4 Rough draft of the system. The first step is to position the patchbays and speakers. Finally, draw in the set of numbered lines at the side and the bottom of the sketch. The grid for the final drawing will be 55 wide and 40 high.

underlay for the final layout. It is a good idea to identify the center of the drawing and the grid, and start from there; that way the diagram will stay centered on the page if extra space is added between equipment groups. The components and wiring can then be drawn on a fresh sheet of paper where the grid indicates. Even at this stage, there is great flexibility in locating components and lines. The grid is a guide, not a rule. The drawing itself never has to be traced, unless you want an inked copy, as the grid positions everything for you.

Finally, draw a rectangle at the bottom right corner of the page for the title and date of the drawing. This position for the title box will be understood when you have to fold a drawing to make it fit in a file or manual. Drawings are zig zag folded both ways with the title on top.

A few short years ago, this was about the only way to set up a drawing. Now, there are computers with simple **Bitmap** drawing programs, more complex but infinitely more flexible **Vector** or **Line Art** programs, and budget and professional **CAD** (computer aided design) programs, all with grids that do not print and can be sized to suit. However, a **roll ruler** for letter size paper, or a small drawing board and a T-square, are portable, fast, and infinitely cheaper. Stencils of rectangles, triangles and circles are available at minimal cost, and a pencil drawing photocopies quite well, looking even better if reduced from a larger drawing. 2B pencils draw dark, stay sharp, and rub out easily.

The final drawing, Fig. 14.5, was done directly from the grid numbers with the most basic of drawing tools; pencil, roll ruler, and A3 copier paper. As in Fig. 14.1, the lettering was typewritten, cut out, and glued on to the drawing, but it satisfied the supervising engineer, and was the drawing framed and displayed next to the equipment rack. A drawing done by the grid method is very readable, and contains as much information as many drawings done in a drafting office, with very small details and lettering. When the drawing of Fig. 14.5 was reduced to A4, no details disappeared, a pointer to diagram component size, which should survive reduction and still convey the same message, like a corporate logo.

PATCHBAY LAYOUT AND WIRING

While microphone, video, and speaker patchbays may use standard connectors that have no provision for normalling, line level patching almost exclusively employs quarter inch (6.3 mm) tip, ring, and sleeve, telephone

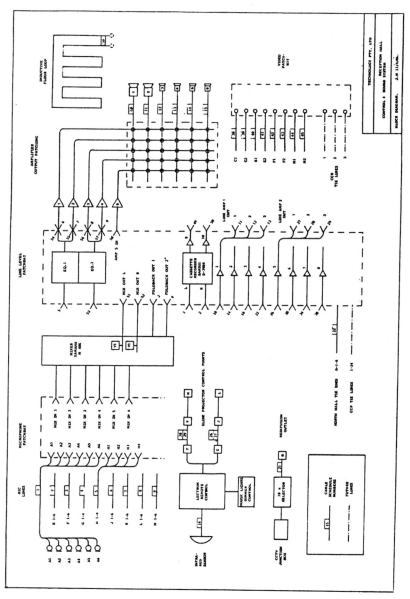

Fig. 14.5 The final drawing, based on the rough sketch of Fig. 14.4. There is a great amount of freedom to vary the original grid numbers and layout.

type jacks, or the 4 mm Bantam jacks, which are the same thing, but occupy half the space. Each row of a 19" jack strip accommodates twenty four standard jacks, or forty eight bantam jacks, but the compact nature of bantams usually requires them to be wired to a separate automatic-terminal strip for normalling and connecting to external lines.

Just as the system drawing flows from left to right, so connections to system elements flow left to right and top to bottom from the front of the jack-field. This arrangement is by no means mandatory; patchbay design is the most flexible of any system that uses standard components, but most operators are used to the standard form where normalling occurs; to the right, or downward. Referring to Figs. 14.2 and 14.6, it will be seen that in most cases the outputs are on the top row, and normal vertically both ways to the inputs on the bottom. Arrows indicate the normals.

The two multiples shown in Fig. 14.2 are parallel groups of jacks for linking up to four lines, and are a very useful facility which should be included even if they are not asked for, if there are spare jacks. The normalling contacts are not used in this case. Sometimes the multiples are included in the monitor selector, so that external portable equipment can be accessed by the monitor speaker without going through the main system.

Some multiples include a terminating resistor which can be patched out by jacking into the marked socket, using the normalling contacts of the jack. The purpose of a patchbay dictates its design. More flexible than a row of switches or a pushbutton matrix, it gives the operator complete control over the program path, but with all jacks removed, a fully normalled jackfield defaults to the most usual setup.

The detail drawing of a Normalled Jack Pair Circuit in Fig 14.3 shows the use of normal contacts to enable access to either line, while breaking any other connection. This configuration usually has a two headed arrow on the designation strip. Other less frequently used configurations enable a jack to be inserted in one line without breaking the connection, indicated by a single headed arrow. Patchbay designations are hand lettered or typed rather than engraved to make them easy to change.

RACKS AND PANELS

Whether or not you make your own equipment racks, it is important to be completely familiar with the dimensions of standard racks, panels and rack-mountable equipment.

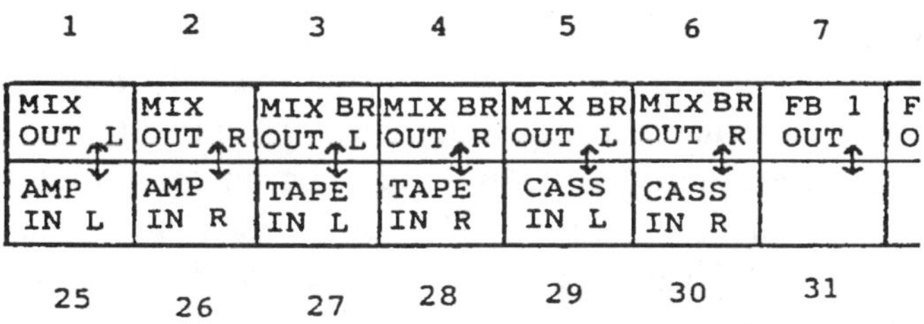

1	2	3	4	5	6	7	
MIX OUT L	MIX OUT R	MIX BR OUT L	MIX BR OUT R	MIX BR OUT L	MIX BR OUT R	FB 1 OUT	F O
AMP IN L	AMP IN R	TAPE IN L	TAPE IN R	CASS IN L	CASS IN R		

| 25 | 26 | 27 | 28 | 29 | 30 | 31 | |

Fig. 14.6 Section of patchbay designation strip showing vertical bi-directional normalling. Single row designations can be attached above and below the jacks on each jack strip, or, as in this case, the double row designation is mounted on a separate panel above the double row jack strips so that it won't be hidden by patchcords.

The **rack unit**, or panel module nineteen inches wide, and one and three quarters of an inch in height, is the basis for all other rack dimensions. Equipment height is in modules of the rack unit, so an equalizer or a double-row jack strip might be 1 RU, an amplifier 2, 3, or 4 RU, and an open reel tape deck or mixer, up to 10 RU.

Blank panels from 1 to 10 RU are used to fill up the rack when all the equipment has been fitted, and also to make up custom items like microphone and speaker patchbays. The table on page 272 shows panel height increments related to the total, and indicates some of the commonly used sizes.

When mounting holes are drilled in either panels or rack mounting rails, the hole centers must be precise, and measured from one end, not from hole to hole, as small errors add together and risk large panels not lining up with the mounting holes. The answer is to make a template; a full length strip template, or a sheet of steel at least the size of the biggest likely panel, say, 19" x 17.5," fully drilled with one eighth inch pilot holes, and checked not only for distance between the holes in each row, but also verified diagonally, so that the two rows of holes are exactly opposite each other, which guarantees that very big panels will fit the rack, and also that smaller panels will mount precisely horizontal.

Once made, the template will be used for marking out panels and aligning the rack bolt-rails, which are best drilled and tapped for the chosen screw thread before assembly on the rack. The rails are then placed **behind** clearance holes in the front rack flanges before welding or countersink bolting on so that mounted equipment will pull the assembly tighter together. If the rails were fixed in front of the sheet metal member, all the weight of the equipment, which can be considerable, would be hanging on the welds or the screws securing the rails to the rack. Tack welds are not particularly strong. Alternatively, captive nuts are available to fit either square or round holes, depending on the source. These are both very good methods because they use oversize nut mounting holes which have enough "jiggle factor" to allow minor drilling and alignment errors to be corrected, and the pre-punched rails are available commercially. The mounting rails are fixed to the rack with the template bolted in place through the four corner holes. Done this way, everything will line up even when large panels are presented to the rack.

In audio installations, which sometimes involve hanging heavy speakers or mounting amplifiers which might each weigh sixty pounds or more, it pays to be constantly aware of the consequences of something coming loose. The same principle applies to mounting amplifiers by the back as well as the front; it's a very simple thing to fix a couple of pre-drilled rails at the back of the rack for supporting the back of heavy amplifiers, as to let them hang on the front flanges only will result in damage to the rack and the amplifiers. Pre-punched steel angle-rail is available which can double as the cable tie-rail. Slide-mounts that bolt in front and back are available for extra-heavy amplifiers. The distance between amplifier front and back mounting holes is not fully standardized (some manufacturers use 15-3/8 inches) but a satisfactory method is to adapt the amplifier to the back support with a short rigid aluminum angle bracket.

Rack Units	Incremental Panel Height	
	Inches	Millimeters
1	1.75	44
2	3.5	89
3	5.25	133
4	7	179
5	8.75	222
6	10.5	267
7	12.25	311
8	14	356
9	15.75	400
10	17.5	445

Safety has to be planned into racks to avoid the possibility of them falling over. If there is any doubt about the stability of a rack, it should have a four or six inch foot on the front of the base. A standard rack 6 feet high or 1800 mm will support 40 rack units, a half rack; 20 R.U. Racks can be made to any height, and the equipment mounted to make the main function controls conveniently accessible, with fixed function or blank filler panels above and below the equipment. Heavy amplifiers should be mounted at the bottom of the rack to ensure stability through a low center of gravity.

Small panels have their uses, taking the place of future equipment, or spacing out amplifiers that depend entirely on convection cooling. Temperature rise specifications of transformers, amplifiers, and the like, pre-suppose free convection air-flow, so it is necessary to make provision for ventilation panels at the top and bottom of enclosed racks, because there is never any certainty what equipment might be fitted at a later date that might generate a lot of heat. Back door and top mounted extractor fans are an option that will ensure that reliability of the system will be greatly extended. If a door is to be fitted, a pair of extractor fans at the top of the rack will be complimented by similar areas of passive vent at the bottom.

Some amplifiers, notably those with MOSFET output devices dissipate extra heat because they need more bias current to offset the inherent crossover distortion due to non-linearity near cutoff. But just because they can run hotter, it doesn't mean they like it; so assisted ventilation is in order if the amplifier relies on convection for cooling. Output transistors running at high temperatures can become loose due to screw expansion and deterioration of the heat transfer compound. When they get loose, heat sink contact fails and they avalanche into failure. MOSFET amplifiers are capable of excellent results, especially in terms of low Transient Intermodulation Distortion, and active ventilation is a good idea if they are likely to be used.

A ventilation system is like an electrical circuit; provided there is no leakage, air flow is the same at any point in the circuit; ten liters a minute at the fan anticipates ten liters a minute past the cooling fins of the heat sinks, but an open back door is a by-pass which will reduce air flow in the heated areas. Just stirring up the air in a rack has no value; the air must flow through it to be effective (Fig. 9.11). If a top mounted extractor fan is used, a back door with a gap or a vent at the bottom is necessary to ensure adequate airflow past all the equipment, and should always blow upward, in the direction of natural convection. It should be verified, in the case of amplifiers with their own fan, that they pull air in at the front and exhaust at the sides or back. If the rack fan moves air the wrong way, a static air condition will result, shortening equipment life.

Power distribution outlets can be either mounted on a panel at the bottom of the rack, or run up the side at the back of the rack, so that each unit has its own power socket. Equipment power cables should be bunched and tied, not cut short. Equipment is removable for service, so the power cords are not fully laced into existing cable looms.

Racks can be designed in both inches and millimeters. Use of dual standards is not recommended in many trades, but since many of the standards in the film, broadcast, and recording industries originated in both inches and millimeters, it pays to be familiar with both. An inch is 25.4 millimeters, and a sixteenth of an inch is about 1.6 mm. With a calculator at hand, rapid conversion between standards is easy.

Whenever metric quantities are chosen, it is recommended that only millimeters should be used for all dimensions of major equipment, other than the details on engraved panels. Meters are for land measurements, and centimeters are for body and clothing measurements. Industries that build trucks or houses in metric countries measure structures in millimeters, including the overall length of the vehicle or building construction. The way is left open for error if meters and fractions of meters are used. It is better communication to say 1087 millimeters than one point 087 meters, or 1 meter and 8.7 centimeters.

ONE INCH EQUALS 25.4 MILLIMETERS.

WORK IN THE STANDARD IN WHICH THE SYSTEM ORIGINATED.

USE MILLIMETERS ONLY, NEVER METERS OR CENTIMETERS, IN CONSTRUCTION PLANS.

USE WHOLE MILLIMETERS FOR PANEL DIMENSIONS. AVOID FRACTIONS.

Screw standards for fixing panels to racks include 3/16th of an inch, and M6 (6 mm). Quarter inch is an older standard.

Steel is the best material for racks, either 1/16th of an inch (1.6 mm) for a lightweight rack, or 2 mm for heavy duty. Left to the choice of the sheet metal fabricator, black iron or mild steel will bend and weld without difficulty, but if welding it yourself, remember that rolled sheet steel shrinks when heated, and if you don't have a pilot arc welder, tack it together first before completing the welds to prevent it bending out of shape.

Make the rack especially strong at the bottom, because the total weight it carries may be considerable (never use pop-rivets to assemble the bottom of the rack; they pop out), and don't forget that some equipment is very much deeper than most audio modules; the rack drawing in Fig. 14.7 may need to be made deeper to accommodate it.

Aluminum can be used for lightweight or low budget one-off racks, but don't ask it to carry much weight if it consists of extrusions and plastic corners. The weight of an amplifier can bend the bolt rail and its supports. If you have to use these components, hide a self-tapping screw inside each leg of every corner casting to lock it, as they tend to work loose. Knock-together aluminum kit racks will not be satisfactory for heavy amplifiers or mobile equipment.

If a rack has removable snap-on sides, installations are very much easier, but this poses some problems for the do-it-yourself constructor. Budget considerations will need to be weighed against the requirement for mechanical strength. If the job warrants it, never hesitate to specify professionally made steel racks with snap-on sides and fully drilled and tapped bolt rails. Depending on your location, a really well made rack with the convenience of removable sides may be an economical choice. In multiple-rack installations, sides between adjacent racks can be left off, conserving cost, and greatly improving access. There are modular assembly rack parts available in most places.

The standard 19" rack drawing in Fig. 14.7 was used for several installations with either one or two racks. In some cases the time period between installation go-ahead and the completion deadline was very short, yet the sheet metal company who constructed the racks at short notice made no mistakes and came through without complaint or query.

In addition to the rack drawing, a chart like the one in Fig 14.8, describing the relationship between mounting holes and panel dimensions should be presented so that the fabricator fully understands the rack system. Unless the constructor is completely familiar with 19" racks, personal liaison is also a good idea before and during the job.

Just as a cable schedule spins off from the block diagram, so rack layout should also be purpose planned from the diagram, even if it's only a freehand sketch (Fig. 14.10).

A cabling floor or plinth will be needed for most fixed installations, similar to the one shown in Fig. 14.9. Cable access both during and after installation needs to be considered, especially if the system might be upgraded later. Figs. 14.9–14.11 feature a two rack installation, showing the plinth, rack positioning relative to work-space back and front of the rack, the equipment layout, and the finished installation. As it turned out, a third rack was added later to the left of the pair, indicating that it pays to be generous when allocating floor space to equipment racks.

EXAMPLES OF STANDARD AND SPECIAL RACKS

The next few pages present practical rack systems and fabrication drawings. Coordinating the site, cable runs, rack bases, and making the best use of the available space are among the factors that warrant committing everything to paper. The job specification often considers none of these things, they are left to the project manager or the installer.

The illustrations show racks that can be made to order by sheet-metal workshops, and also the simplicity of site-plan drawings that will satisfy project engineers as well as serve to brief construction crews. Remember that builders work strictly to the dimensions given on a drawing, and will not stop to measure your scale drawing if written dimensions are incorrect.

A set of sketches, easily converted into scale drawings like the site plan and plinth detail in Fig. 14.9, will eliminate the syndrome: "There's never time to do it right, but there's always time to do it twice."

Drawings can be finished at home in a leisurely fashion. Time spent on the site is more expensive than time spent at the workshop pre-wiring and drilling racks, mounting all but the heaviest rack equipment, and making it all fit before transporting it to the job. Racks and panels don't have to be do-it-yourself, but the following illustrations will serve to familiarize the reader with many phases of rack design and placement, so that commercial racks also fit better into a project.

Again, beware of purchasing panels and panel fronted equipment boxes from hobby stores. They look professional, but they may have non-standard heights.

COMMUNICATING IDEAS

An advantage in being familiar with rack dimensions and reasonably adept at drawing places the installer or planning technician far ahead in ability to *communicate* the details to other people.

Project managers, supervising engineers, construction foremen, your own staff, builders, and rack and panel fabricators; all these people can read drawings and will instantly get the message if one is presented.

Discussion exchanges ideas. Drawings lock them in place.

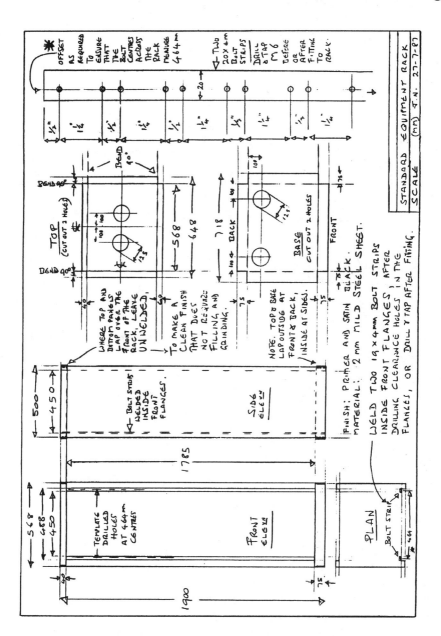

Fig. 14.7 This drawing of a 40 RU (rack units) equipment rack has been used to construct numerous full size racks, about 6 feet high, and also a number of 20 RU half racks.

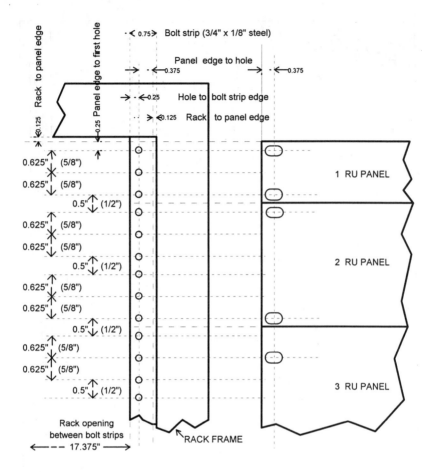

Fig. 14.8 Relationship between the mounting rails on the rack, and various size panels.

Rack hole positions allow a 1/8" margin all round the panels, then start with a 1/4" space (half a 1/2" space), and then repeat the sequence: 5/8", 5/8", 1/2", to the end of the bolt rail, finishing again in a 1/4" space.

Whereas 1 and 2 RU (rack units) panels generally use the hole nearest the top and bottom edges, larger panels start in 7/8 of an inch or more, to spread the load and make the panel look better. On panels over 4 RU, screw holes are repeated at 2 or 3 RU intervals.

If captive nuts are used, square holes follow the same sequence, but the captive nuts for round 5/16" clearance holes permit enough adjustment to equally space all the holes at the mean distance of approximately 0.58".

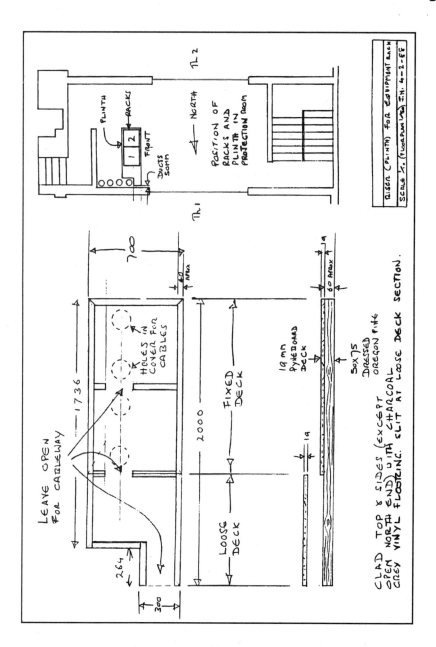

Fig. 14.9 Site plan showing plinth, cableways, and positon of two standard racks. The racks and equipment layouts are shown in Figs. 14.10 and 14.11.

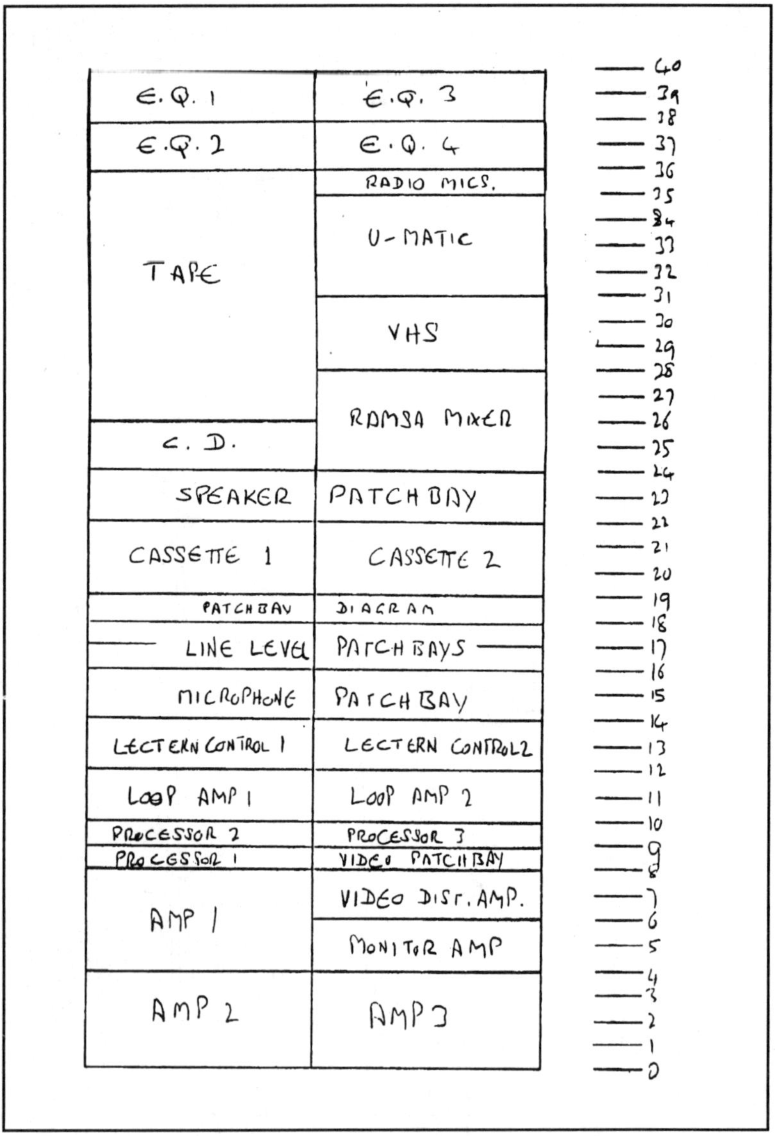

E. Q. 1	E. Q. 3	— 40
		— 39
		— 38
E. Q. 2	E. Q. 4	— 37
		— 36
	RADIO MICS.	— 35
		— 34
	U-MATIC	— 33
TAPE		— 32
		— 31
	VHS	— 30
		— 29
		— 28
		— 27
	RAMSA MIXER	— 26
C. D.		— 25
		— 24
SPEAKER	PATCH BAY	— 23
		— 22
CASSETTE 1	CASSETTE 2	— 21
		— 20
PATCH BAY	DIAGRAM	— 19
		— 18
— LINE LEVEL	PATCH BAYS —	— 17
		— 16
MICROPHONE	PATCH BAY	— 15
		— 14
LECTERN CONTROL 1	LECTERN CONTROL 2	— 13
		— 12
LOOP AMP 1	LOOP AMP 2	— 11
		— 10
PROCESSOR 2	PROCESSOR 3	— 9
PROCESSOR 1	VIDEO PATCH BAY	— 8
	VIDEO DIST. AMP.	— 7
AMP 1		— 6
	MONITOR AMP	— 5
		— 4
AMP 2	AMP 3	— 3
		— 2
		— 1
		— 0

Fig. 14.10 Rack layout prepared from the block diagram of two Audio Visual Theaters. The completed racks are shown in the photograph of Fig. 14.11.

Fig. 14.11 Finished racks shown in site plan (Fig. 14.9).

FRONT OPENING RACKS

It is customary to position racks so that there is a work-space behind them, as shown in Fig.14.9, but very often in confined areas they have to be pushed up against the wall to make enough room at the front. While creating an inaccessible rack in this manner is to be avoided, the layout of technical areas is sometimes out of our hands, and further work is done on the racks by moving them around. However, if the metalwork is being made to order, it may be appropriate to specify a front-opening rack, made to your own design. Commissioning a specialist manufacturer to produce it will cost almost twice as much, because **creativity**, other than your own, is the most expensive commodity available.

In the specification for the rack in Fig. 14.2, five-way cable segregation was called for; Microphone, Line-level audio, Speakers, Video, and Power and control. The top, bottom, and sides of the cable tie frame mounted behind the hinged door provided separate paths for the different types to every piece of equipment. The cables outside the rack were all enclosed in screwed conduit as specified, entering through the separate gland plate which was fully fitted before the rack was moved into place.

What happens to the cables when the door rotates 90 degrees? There are two ways to make a flexible cable section around a hinge. Either a vertical two foot cable length laid alongside the hinge axis twists with the door movement, or in this case, because cable groups had to remain separated, free loops representing each cable group formed a horizontal 'U' about six inches in diameter, one above the other, secured in the rack body by tie points, and on the door, by the segregation frame.

By choice, the heavier equipment was located nearest the hinge. Single racks don't lend themselves to front opening without double-jointed hinges or a distance piece to give clearance to the back of the equipment as it swings open, but a six inch wide panel was adequate in this case.

PANEL DESIGN

Most rack mounted equipment comes from a manufacturer, but in almost every installation there are some panels you have to make or commission yourself. Whichever applies, the layout for active and blank panels should first be drawn.

There are exceptions, like the monitor selector switch and gain control that can be made up on site with a 3/8" drill, some stick-on lettering and

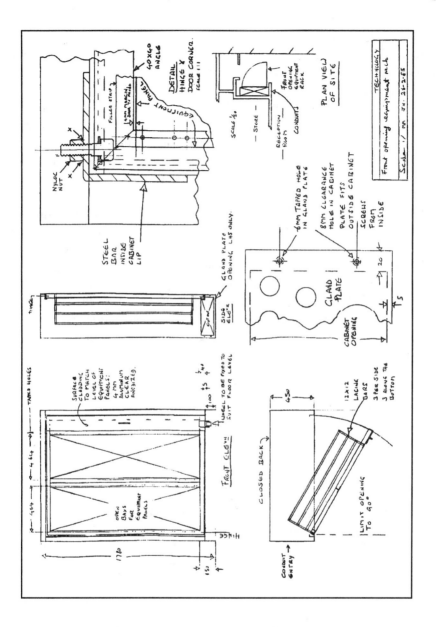

Fig. 14.12 This front opening variation of the standard rack has two bays totaling 80 RU. The panel mounting holes are drilled and the threads tapped directly into the angle-iron door frame.

a piece of clear acrylic sheet to cover the designations. But the principle of drawing first, then acting, is never more appropriate than for one panel, or a series of panels you make yourself.

The standard heights of 19" panels are given on page 272, and the position of the mounting holes in Fig. 14.8. Every panel is made short of the nominal height by a thirty-second of an inch all round. This allows for paint thickness and cutting error, so that there will not be a net height gain caused by small errors adding in the same direction. It also provides a degree of "jiggle-factor" which makes up for drilling and other inaccuracies.

The panel width is also critical. It can not afford to be exactly nineteen inches, but just under, so that in the rare but occasional event of teaming up with a rack whose panel opening is precisely nineteen inches, there will not have to be a choice between an interference fit or using a file on the panel. Conversely, I've seen quite expensive equipment like mixers and tape decks on panels that are more than 19 inches wide, so if you are drawing up a rack with sides enclosing the panel, add a quarter of an inch to the opening width to guarantee clearance under all conditions.

The electronic trades have long been accused of drilling five thirty second holes for one eighth screws, but in situations where one manufacturer makes the rack, and others makes the panels, the concept of **room for small errors** is valid. Most panels, for the same reason, have slotted mounting holes.

CONTROL GROUPING

It's likely that you have used infra-red remote controls of various brands that consisted of a regular button matrix, each control labelled in identical style, and on which the only way to locate a function in semi-darkness is to count from the top, and then from the left.

Other manufacturers approach the situation in a more oblique manner, grouping the main controls in specific areas, and perhaps using buttons of different sizes and shapes to form recognizable patterns. The result is a **control grouped** design that is easy to read; much like the open style block diagrams described earlier.

Control grouping deserves first mention when discussing panel layouts, because it is fundamental to user-friendly design, and differentiates the "real world" side of a control panel from the wiring side which merely needs to contain neat rows of components. Control grouping, which

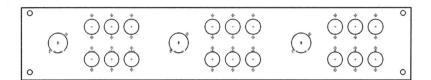

Fig. 14.13a A 19" panel layout, ready for drilling.

This panel is finished, the holes are the right size, and the screw spacing is correct.

Q. What is wrong with it?

A. There is no room for designations. The connectors need to be offset so they can be labelled or engraved.

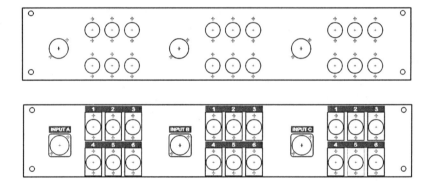

Fig. 14.13b The panel layout corrected and labelled with the connector profiles overlaid. It's easily forgotten when you're concentrating on getting the holes right for the connectors, but if you only *sketch* the panel first, it will immediately be apparent that it costs no more to offset the holes so that the labels can be applied later.

sometimes includes enclosing lines drawn around the groups, enables operators to locate the item they are looking for at a glance, because the recognition pattern of the control group is unique, and leads them straight to it.

The principle applies equally to connector panels like the ones shown in Figs. 14.13a and b. Remote system panels like the one in Fig. 14.14 are often located in dark corners or under floor-traps, and all efforts should be made to make the job easier to read. Panels weakened by large holes should be of heavier material, or strengthened on the back with a rib of aluminum angle.

Observing the style of various pieces of commercial equipment will give you the direction to take with your panel layouts.

Applying the principle of control grouping gives panels you make a professional appearance. Intelligent panel layout makes all the difference between a rack that looks home-made, and one where the custom made panels match major equipment like mixers and tape decks.

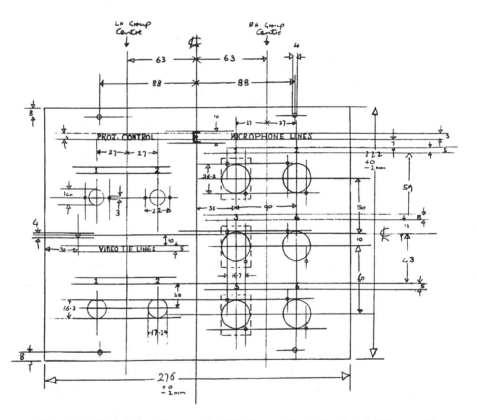

Fig. 14.14 Panel drawing marked "E," one of a series for drilling and engraving. The original was full size on tracing paper. Note that the centers of the left and right hand connector groups are referenced to the Center Line reference points.

LAYING OUT A PANEL

I don't know if you have ever experienced that feeling at the start of a job that leaves you wondering just where to begin. One of the best launch vehicles for a design "brainstorm" is a scribbled drawing of how the finished article will look. It doesn't matter how rough it is; probably nobody but you will ever see it, but it will trigger the thinking process. It's part of the natural ability we all have, in which the process of **visualizing** the finished job programs the mind with enough data to get us started.

Just as a block diagram starts with a rectangle representing the line level patchbay, so also a series of squares drawn to position the main control groups commences the panel design, and then the mechanics of filling in the control group areas and arranging the spaces between components follows automatically.

If the finished panel is seen first in the mind's eye, then our inherent abilities take over. Without first visualizing the finished panel before pursuing task of laying out controls, the panel will remain exactly what it is; a piece of metal with a pattern on it. But visualize it, draw in the groups, allowing the subconscious mind to effect the design, and the whole becomes greater than the sum of it's parts.

Never consider for a moment that this ability requires special talent or training; we all have it, and it just needs a starting point and a few guidelines.

LETTERING, STYLING, AND ENGRAVING

It will be noticed that the panel in Fig. 14.14 has fairly large lettering. This is OK under the dark corner conditions expected of a remote connector panel, but it should be noted that the lettering size on most rack mounted operator panels is quite small. 5 mm high letters on a rack panel would be as out of place as sign-writing. To get an idea of acceptable lettering heights, a good exercise is to look at some commercial equipment first.

Measure the letter heights, because they are not what they appear to be. It will be noticed that the average height for lettering is around 3/32" (2.4 mm), or 1/16" (1.6 mm).

Something else that will be noticed on well designed commercially made panels is that the letter size varies, and that minor lettering is much smaller than would be expected. Some lettering might appear to be an eighth of an inch high, but if you ordered that size from an engraver for one of your own panels, the result would look more like 3/16". This is partly because, unless otherwise directed, engravers make the specified size **the center of the cut**. If the line thickness is for **Bold letters**, the resultant lettering will be still larger. So if in doubt, order 3/32" lettering for a finished size of 1/8". A safe way to brief an engraver is to present a sample of some other work, and remember that if you order a style with sharp, square corners, like Helvetica, the letters will finish closer to specified size, and the cost will be higher, because square corner engraving involves making more passes over the job with finer tool-tips.

Engraving looks best paint-filled. This works better than unfilled letters because the latter tend fill with dirt, and the base metal tarnishes with age. Deep-cut unfilled letter engraving loses readability if viewed at an angle. One of the best looking methods is engraving a clear acrylic overlay on the back in reverse, paint-filling the letters in one or more colors, and then spraying the background color. Depending on the quantity of each panel required, the panel can be silkscreen printed in reverse on the back, resulting in considerable cost saving. When specifying acrylic overlays, have them cut slightly smaller than the metal backing panel, to protect the edges from catching and breaking. When ordering plastic panels, either for reverse engraving or two-color laminates for front engraving, specify **Do Not Bevel Edges**, unless that is your requirement, because some engravers will do it as a matter of course.

When you draw a panel for engraving, double check everything, because a labor intensive and therefore expensive process follows. Leave nothing to chance; whenever you present a drawing, mark all dimensions, and present it just as it will finish; actual size, mirror image if it is for reverse engraving, with full instructions in writing as well as the drawing.

SCALES FOR POTENTIOMETERS AND SWITCHES

It is always advisable to double-check with the actual components you intend to use, but most potentiometers have 270 degrees of travel, or three quadrants, and rotary switches are designed around twelve positions, which makes each division 30 degrees. Like the scales in Fig. 14.15, use any styling trick you can to make the scales lighter and more attractive.

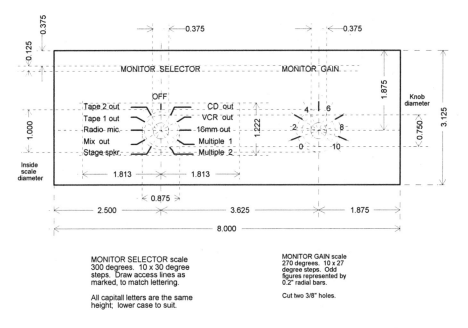

Fig. 14.15 Typical scales for potentiometers and rotary switches, showing constructions. The odd number bars on the gain scale are an option which makes the scale look lighter. Include all dimensions and lines when making drawings for the engraver. Note that the engraved overlay should be slightly smaller than rack panel height. Blunt sharp edges with fine sandpaper on a wood-block. The source designations can alternatively be clear windows for hand lettered inserts.

BEFORE STARTING WORK ON A PANEL

There are two essentials of panel design that have to be considered before starting. As Fig. 14.13 indicates, the layout needs to include labels, or at least a space for them.

Second, double-check the gender and brand of the connectors. There is no greater waste of time and energy than remaking a panel because the big holes should be small and the small holes big, but it can happen to anyone who has not made it a habit to always double-check.

It will also be noticed that not all makes or styles of the same connector use identical holes. For example, some BNC connectors use round holes, and some have a flat on one side. It is often better to have odd shaped holes cut by an engraver, unless you're really fast and accurate with a file.

MATERIALS AND PANEL THICKNESS

Both steel and aluminum are good materials for panels, but if they are to be worked with a band-saw, hole-saw, or drill, there's a world of difference between steels, and between different grades of aluminum.

Mild steel plate should be chosen for workability and flatness. Boiler plate is not flat, black-iron is irregularly surface-hard, will blunt your drills, and does not bend perfectly. Steel should be primed before paint finishing; don't rely on one-coat paints unless you are sure of their performance, as steel will rust even if the paint stays on.

Avoid **stainless steel** unless it is specified or destined for a corrosive environment, because it fingerprints badly, and it's more expensive. If it is specified, let the engraver cut the large holes, stainless is tricky if you are not familiar with it.

Aluminum can be anodized in any color; black or clear can be used to match other equipment. **Hard plate** is best for anodizing, as some soft grades finish badly. Engrave first, as anodizing forms a thin layer of sapphire (aluminum oxide) on the surface and will blunt the cutters and shatter the edges of the cut. Care has to be taken paint-filling engraving, as the anodizing can stain. Castings don't anodize well.

Aluminum sheet for plain anodizing has to be selected and handled carefully to avoid marking, but a brushed finish is very practical. Some soft aluminum sheet is difficult to drill and tap, it's not always flat, and marks badly. Panels call for a flat sheet with hard edges that cuts cleanly and doesn't clog the drills. Dropping a piece of soft aluminum on the floor often causes corner damage that can't be fixed. Hard plate drills and taps well, but can not be bent. If in doubt, ask the anodizer to suggest a grade, or conduct a test before buying the batch.

In all panel materials, choose a gauge that will not bend seriously with the pressure of a connector being pushed in or extracted. In a full rack width panel, an eighth of an inch for steel, and three sixteenths for aluminum, are safe minimum thicknesses, but the gauge goes up if the panel is seriously weakened by large holes. Filler panels can be lighter, or even made of 16 gauge sheet, bent 90 degrees for up to 17-1/2 inches of the long sides to provide rigidity. Blank panels are available with slotted rack mounting holes already punched or milled.

If a large panel is weakened by too many holes, which is sometimes unavoidable, a piece of aluminum angle bolted to the back will make a deep web to stiffen the panel at convenient places. Bear in mind that the rack opening may not be any more than 17-1/2 inches, and that components, bends, and webs should be kept within that margin. Strengthening webs can be fixed under connector screws to save making extra holes.

PANELS FOR MEMBRANE SWITCHES

A membrane switch panel presents a flat surface, sealed by a flexible polyester overlay, covering either printed circuit switch contacts or micro-displacement switches. The overlay is silk screen printed on the back with the switch functions and other control group information.

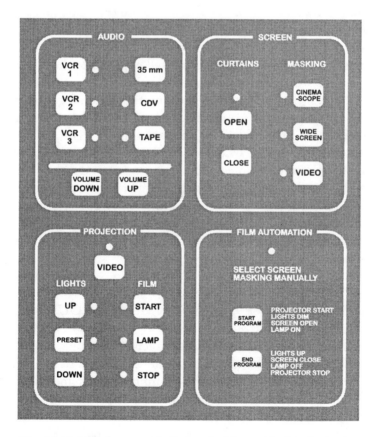

Fig. 14.16 Membrane switch panel. Alternatively, the source designations can have clear windows for hand lettered inserts.

Two such panels were used in an AV Theaterette; one 6-3/4 x 8 inch remote panel in the theater as shown in Fig. 14.16, and a 19" x 3RU panel with the four groups deployed horizontally on the main equipment rack in the projection room.

Micro-displacement switches are very reliable, so the likelihood of having to dismantle the panel assembly is remote. They are generally available only in a single momentary "make" contact, and are not designed to carry appreciable current, so a relay or solid state latching circuit is used. The program source selector and film automation relay circuits illustrated in Chapter 12 use make contacts for all functions, and were used with the membrane switch panels described above.

One advantage of using only normally open, or "make" contacts, is that switch circuits then have greater inherent reliability, as resistance that might develop in a normally closed contact set is eliminated as a failure factor. Like many techniques, the decision to use this one usually depends on how many you intend to make, as membrane switch panels are economical in quantities over about ten. However, particular contracts that will be advantaged by the sophisticated appearance that the method presents, will be worth making in smaller quantities, even one or two.

Making the panel is a matter of selecting the material gauge, plus the thickness of a sub-panel, to match the depth of the available switches. The composite panel is then backed by an insulating layer strong enough to handle the pressure on the switches. The leads are brought through the insulating support, and a printed circuit or matrix board is then secured to the assembly and the connections are made. The screen-printed polyester overlay is then fixed to the panel using double-sided adhesive film.

Optional clear sections can be left for LED indicators, and to permit alternative designations to be inserted under the panel on thin adhesive polyester tape.

BRACKETS AND SHELF UNITS FOR MOBILE EQUIPMENT

Most equipment like VCRs and Cassette Recorders can be rack mounted by fitting right angle brackets to the sides under the case screws to mount them on the rack's bolt rails. The procedure necessarily means that each item will need custom made brackets, because there is little size uniformity among freestanding or portable equipment.

But it is often desirable to place these units in a rack without bolting them in, or so they can be readily changed or moved from place to place.

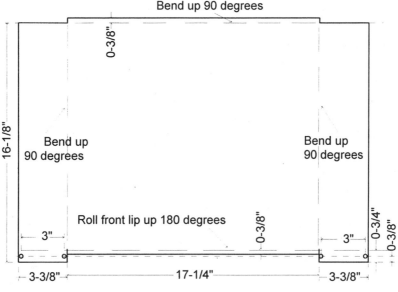

Bend up 90 degrees

Bend bolt flanges out 90 degrees

Fig. 14.17 Construction of rack shelf unit.

Shelf units are available from some audio suppliers which offer the maximum width and depth of the rack, and suit two, three, or four rack unit heights. Alternatively, as shown in Fig. 14.17, it is a very simple matter to cut and bend them to order from 16 gauge aluminum or 18 gauge steel, allowing a maximum internal width of 17-1/4 inches, which will fit into a standard rack and accommodates most equipment that has no mounting flanges. Note that marking the drawing in fractions rather than tenths is often more helpful when working with a rack system based on eighth of an inch dimension modules.

The shelf unit height will determine the load it is capable of supporting, but most source equipment is relatively light, and if the mounting flanges, for example, are two rack units high, which is the recommended minimum, the shelf will support the weight of most domestic VCRs. The lip on the back stops things sliding off, and the front lip, which is loosely folded back on itself, gives the shelf strength.

DRILLING PRECISION HOLES IN METAL AND PLASTIC

No discussion on panels is complete without a few pointers on how to make holes in them, for even in audio, there are difficult jobs requiring precision centering and assembly, and also some materials that shatter and split when approached with an ordinary drill. One of the main difficulties is the tendency of drills to drift sideways, and as progressively larger drills are used to make big holes, the drift and grab becomes worse, and then the drill orbits around true center resulting in a three sided hole instead of a round one, and/or shattering the edge of the cut.

The solution to all these troubles is the use of shaped drills. As shown in Fig. 14.18, the first maneuver is to grind away the cutting edge to form a centering tip. When this is done, center-punch the panel and use a pilot drill small enough to made a tight hole for the guide tip. To maintain center, the full size drill is used immediately, without intermediate steps. The hole stays centered, and holes up to at least 7/8" diameter can be cut with perfectly clean sides.

This method is good in aluminum and mild steel. Steel needs oil as a lubricant. I won't advocate it for stainless steel, as some of it is quite difficult to drill, and all but very slow drills will burn and blunt in large holes, so use a standard drill or have the hole cut by the engraver.

Brass, acrylic plastic sheet, and phenolic resin impregnated pressed paper products need an additional technique. Whether or not the drill has been cut away to form a center guide, these materials need a pair of regulator facets on the drill to stop it grabbing. If the cutting edge of the drill is given a pair of flats 90 degrees to the plane of the work, the cutting speed will be slowed, and the drill will not grab as it breaks through. Drill modifications should be tried on a piece of scrap first. If soft aluminum has to be drilled or tapped, frequent applications of oil will make the job easier.

At all times, the panel should be pressed down firmly on to a flat piece of wood which guides the tip after it penetrates the sheet, and also keeps the outer edge of the drill in solid material as it breaks through. When fitting two connectors that have to be precisely spaced, insert a matching plug into both sockets before tightening the fasteners. Example; sockets for a dual banana plug.

The pressure needed to drill different materials varies. More for hard materials, less for soft. If a drill rotates too long in one place without advancing, it will get hot and the cutting edge will spoil. Plaster-board fasteners represent the most extreme condition anyone is likely to en-

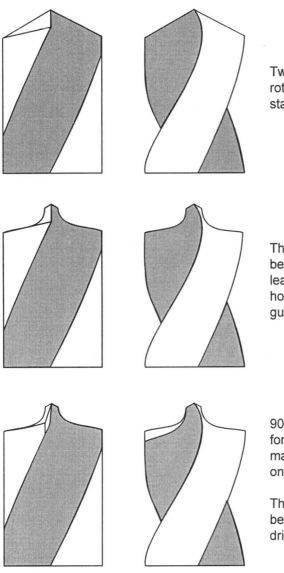

Two views, 90 degrees rotated, showing a standard drill.

The cutting edges have been ground away, leaving them almost horizontal, to form a guide tip.

90 degree cutting facets for difficult materials are made with a light touch on the grinder.

This method can also be used on standard drills.

Fig. 14.18 Drill tips modified for panel work.

counter. The surface material is very soft, and requires restraint to get a close-fitting hole, using a standard drill-tip. Let the drill find its own way in, but be cautious when drilling through folded steel wall studs, as they will bend away from the drill if too much pressure is applied, resulting in an oversize hole that will not grip the fastener.

Chapter 15.

Home Theater Sound

A large portion of this book has concentrated on commercial sound techniques and equipment, but there is every reason for the Audio Professional as well as the Audio Enthusiast to pursue the subject and practice of listening to good sound at home, and to appreciate the extraordinary perfection of available sound sources and processors.

Home Theater is the ultimate extension of listening to TV sound through the stereo system that plays your tapes and compact discs. The concept of **Total Theater** in your home can be executed so well that it lacks only the empathy of a large audience.

Motion Picture Sound is advancing so fast that every two or three years there is a new format, not always backward compatible with earlier systems. These include new speaker systems and multi-channel amplifiers, addition of a sub-bass channel to reproduce the last octave in the spectrum, bi-amplification of high and low frequency speaker components, and noise reduction carried to its limits.

Digital sound has entered the Cinema and also the Home Theater with five discrete channels plus sub-bass. As might be expected, a whole new generation of movies is appearing to take advantage of the wider response and greater dynamic range available.

The basic stereo system which revolutionized cinema in the 1980s is the one still used in most home theater systems. Referring to the experimental surround system shown in Fig. 6.3, it is evident that all stereo pairs can produce a total of three extra tracks by adding and subtracting the two original signals; left plus right, left minus right, and right minus left. Although it states in Chapter 6 that decoders attenuate the surround channel below 90 Hz, there is still a need for full range good quality surround speakers. Movie producers are taking advantage of the extra dimension by including a lot of bass in the surround, independently of the front channels channels, and much of the soundtrack's value is lost if it is not reproduced.

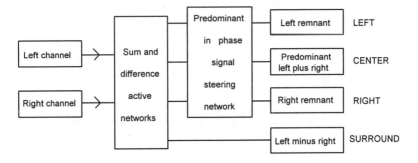

Fig. 15.1 The basic stereo cinema processor derives four channels
from the original two.

The stereo cinema system (Fig. 15.1) produces four discreet channels
that have complete separation, and if necessary can reproduce four
different signals without noticeable interaction. Such is the basis of home
theater sound, shown in a typical system in Fig. 15.2.

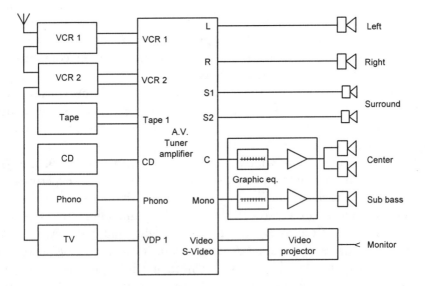

Fig. 15.2 Typical home theater system. Insert points for center
channel processors are not fitted to many decoder-amplifiers, so an
extra power amplifier follows the graphic equalizer, the other side
being used to filter the sub-bass.

REAL THEATER SOUND AT HOME

In the days of monaural television broadcasts, before video recorders were generally available, audiophiles would pipe the TV sound though their mono or stereo Hi Fi systems and gain a worthwhile dimension of quality listening. It would have been noticed that the programs did not all sound the same, and that the purpose bandwidth limitation (not the same as FM high frequency de-emphasis) imposed on TV receiver sound channels was there partly to conceal a wide disparity in the tonal balance of programs.

Reproducing TV programs with a wide frequency range reveals that some are thin, some lack middle frequencies, and others are bassy. Does this indicate there is something wrong with Hi Fi systems, in particular the speakers? No, but it does show that wide range sound needs careful tailoring to **centralize** the response balance, with particular attention to the middle band of frequencies which are crucial to dialog. In fact, many home videophiles install a graphic equalizer in every channel and calibrate their acoustic response exactly as installation and maintenance technicians do in auditoriums and theaters. (Reference Chapter 6: Auditorium and Speaker Calibration).

One of the essential truths in audio is that when frequency response is optimized, half the problems of noise and program balance simply fade away.

Reproduction in a home theater system can sound just as good, or better, than the same program in a big auditorium. The Cinema Acoustic Characteristic (Fig. 6.16) is used in large auditoriums to combat multiple-length reverberation paths (page 125) but high frequency reverberation is far less significant in small rooms. Speakers should be of the highest quality, suitable for music reproduction, and the center channel equalized to achieve a natural dialog sound. If you use two center channel speakers, one on either side of a projection screen, they should be not less than four feet apart. This avoids serious interaction between speakers carrying identical signals (pages 106 and 107).

A single speaker above the screen will also do the job, but if it is near a TV picture tube it must be a magnetically shielded type, or the screen will lose color purity in the area near the speaker, and it will have to be de-magnetized. Sound from a TV receiver's own speaker system can not successfully represent center channel, as it does not reproduce a predominant signal extracted from left and right, and in any event, The TV's own speakers are not likely to match the rest of the Home Theater speaker array.

Center channel carries nearly all the dialog, but it also represents everything that is seen within the picture boundaries, such as main on-screen effects, and just as much music as left and right, especially the main title music. Also, don't forget that some of the great movies are monaural, and they can sound really good if the center channel stands up by itself, because center and sub-bass is all that is operating on a mono program. (Stereo simulators won't decode into more spatially correct channels; they can not create true stereo information). There is provision in most analog home theater decoders for use of a center channel speaker with less than full range bass response. Decoders with this facility have a mode switch that includes NORMAL and WIDE settings.

When using a bass limited center speaker, the NORMAL setting is selected. This steers all the bass to left and right so that none is lost, while the middle and high frequencies are steered to Left, Center, and Right, as directed by the program content. However, the WIDE setting steers all frequencies impartially to the three front speakers, and presupposes that Left, Center, and Right all reproduce full range sound.

If a restricted bass center speaker is used with the WIDE setting, much of the bass content of the program will be lost. The sub-bass channel will not replace it, as it does not usually cover the main low frequency band of the front speakers, and may leave a hole between 50 and 150 Hz. The same comment applies to monaural programs where only the center and sub-bass speakers are operating, and they might sound better reproduced in the stereo mode. A mono program generates no surround information.

Unlike stereo sound, equal power is not propagated by all channels at the same time, although equal power capability is recommended for all channels, including surround and sub-bass. The distribution of soundtrack program peaks is different for movie sound compared to the orderly self-control of music, so 50 watts per channel is minimum recommended amplifier power for every home theater channel in a small room, and 100 watts is recommended for each channel for large rooms. Both the amplifiers and the speakers must be able to handle peaks at all parts of the spectrum without sounding stressed.

SURROUND SPEAKER POSITION

The surround is an ambient channel which benefits from wide dispersal. In theaters, rows of speakers face each other along the side walls, above the listeners heads. Proportionately little surround sound is propagated from the back wall. Depending on the room you use, a good starting point is to face the surround speakers towards each other on the side walls about seven feet high, the same distance back as the listeners.

Surround is not a rear channel, but as the name implies, it should surround the listeners. Various positions can be tried until the surround is satisfactory, although some system manufacturers give more specific placement instructions.

EQUALIZING THE CENTER CHANNEL

With reference to the Dialog Equalization process described in Chapter 8, when soundtracks are assembled, the importance of the right "color" of reproduced sound can not be overstated. Human speech has an intensely recognizable "flavor" for want of a better word, and if it is out of balance, it will be quite obvious to the ear, spoiling the cinematic illusion.

Sound effects too, representing things we know in nature, need to be reproduced in balance. If they don't sound real, it is probably because we are listening to a hi-fi center speaker that sounds great on music, the forgiving program, but needs correction to suit it for the uncompromising sounds of the real world.

The most fragile part of the dialog spectrum is the region which contains the fundamental power frequencies of well balanced speech. It basically ends at 2 kHz. It's true there are speech fundamentals above 2 k, but the band from 1,500 cycles down to 500 is the spectrum that needs truth in reproduction. All sorts of things can be done to the bands outside this region, a falling bass and treble response will effectively bring it up in level, but unless the 500 to 1,500 Hz band effectively is in proportion, the dialog will have unnatural quality, and will be hard to understand. The Case History on the following page illustrates the point.

One case doesn't prove it, but experience will bear witness that the central band of speech should not be bent, and should remain on the centerline of the response chart.

Setting up the center channel graphic equalizer is not as simple as it first appears, but with the foregoing information in mind, it can be done without using a spectrum analyzer. Pick a variety of known good broadcast programs which you have heard in the theater. Ignore low budget programs, and most items in off-peak viewing times, but using broadcasts of relatively recent movies, listen in prime time when the good ones are on. Audition videotapes and laserdiscs of soundtracks that you have heard

CASE HISTORY

BASS AND TREBLE OK, BUT SOUND IS BAD.

VENUE: A Regularly Serviced Club Auditorium.

PROBLEM: Emergency Call ; Dialog Unintelligible.

INVESTIGATION: Pedestal mounted Xenon Arc 16 mm Projector was still in perfect alignment, but treble and bass tone controls were found to have been turned up to maximum.

SERVICE PROCEDURE:

Centralize tone controls, remove knobs, and fit potentiometer locks.

in a theater, with the picture of course, because one's perception of a soundtrack is quite different without it.

TV broadcasts are modified by compression and limiting prior to transmission. Tape and disc releases will therefore sound somewhat different, but other than dynamic range, I have not found any problems that spoil appreciation of most off-air programs.

Remember that it is how it sounds that counts, not the position of the controls, because if the speaker needs correction, then the drive you are applying to it will not be a flat response, but the inverse of the speaker's error. Reducing the bass and treble effectively raises the middle of the spectrum, so cut rather than boost to bring up the dialog presence band.

When setting up the graphic equalizer, make small changes, and repeat the whole alignment several times to refine them. Mark the position of each control with a wax pencil so the provisional setting can be restored following any experimentation, and then listen to quality program for a while, until you are satisfied with the result. Graphic equalizers are a "set and forget" device. They should not be tweaked every time something doesn't sound perfect, because programs vary and so does our perception, but if there is a lingering dissatisfaction with the calibration, it should be tried again later to find a better setting.

It may take a while to establish a satisfactory center channel or dialog sound, but it's worth the trouble. If you recall the closing comments regarding auditorium calibration on page 131, in point 9, it is quite in order to make minor subjective changes to the overall bass and treble balances following a calibration series, because even if you use a spectrum analyzer, it does not tell the entire story, and there are areas where your ears will be a better final judge of dialog quality, especially when confronted with acoustic differences between rooms.

Sound effects also need true reproduction if they are to fit the picture's perspective. A high percentage of on-screen effects come from the center speaker, and they will sound real if the center channel is correctly equalized for dialog. Dialog is an intensely recognizable signal, and the listener instinctively knows when it is in natural balance. If you set up your center channel using a broad sample of the best movie dialog, then all the other programs, including direct broadcasts, will automatically achieve their optimum result. But don't start with these other programs as your guide, because they won't necessarily tell you what good soundtrack dialog should sound like.

Music should not be compromised by center channel dialog alignment, as bandpassing has already been done in the dialog mix, and should not be repeated to excess in the Home Theater. However, at the same time, system noise bands and extreme highs should be adequately contained, because they can spoil an otherwise perfect program. High frequency imbalance is perceived differently by people of different ages. Young people will hear the heterodyne whistles and other noises of broadcast transmission error. Older ears, unable to resolve frequencies above 12 kHz or so, can either fail to detect such noise, or they may hear excessive program high frequencies as noise.

Finally, use the TEST facility on the decoder, if it has one, to set up identical levels on the three front channels. The surround can also be set this way, but the surround content of some movies may call for individual settings. If you are happy with the center channel equalization you have set, mark each slider on the graphic equalizer with a small dot of liquid paper, wax pencil, or something easily removed. Then you can experiment and return to the original setting at any time. If you have one, fit a security cover on the graphic equalizer to keep out wandering fingers.

ADDING A SUB-BASS

The audio spectrum we hear has approximately ten octaves:

1	10,000	to	20,000 Hz
2	5,000	to	10,000 Hz
3	2,500	to	5,000 Hz
4	1,250	to	2,500 Hz
5	625	to	1,250 Hz
6	313	to	625 Hz
7	156	to	313 Hz
8	78	to	156 Hz
9	39	to	78 Hz
10	19.5	to	39 Hz

If we want to hear the best sound possible, it doesn't make a lot of sense to drop the last octave if it can be easily reproduced. Pages 109 and 110 have some comments about the sub-bass channel, but the majority of systems do not offer this facility, and in any event, the need for bass extension applies also to stereo music enjoyment, particularly where acoustic instruments are involved.

Subsonics are a major part of live performances, and if you haven't been to an acoustic instrument concert lately, it will come as a shock to attend one and realize what has been missing from your music. Much of the missing part can be recovered by adding the tenth octave to your system, from 40 Hz down to 20 Hz.

Graphic equalizers, whose response structure is illustrated in Fig. 5.4b, are not ideal, although they can be used, because their action is a series of broad overlapping peaks produced by tuned filters. This means that while a 10 band graphic covers the normal spectrum quite well, there is no guarantee that the 20 Hz region will be particularly well served by the lowest band, centered on 30 Hz, because the intention of using a sub-bass is to cover an extended spectrum. Either a 31 band graphic or a purpose designed filter is going to give a better result.

Such filters are available in the form of a two-input mixer, followed by the filter section, which is designed to bridge the Left and Right speaker outputs of the amplifier and thus sample the combined bass spectrum of both channels. If the front speaker program has been split into left, center, and right by a decoder, then it will have to be set on NORMAL,

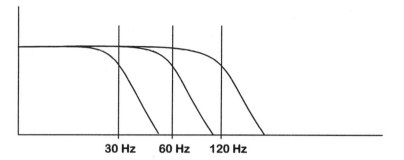

Fig. 15.3 Alternative cut-off points of a variable sub-bass (low-pass) filter.

not WIDE, to capture all the bass (page 300), otherwise, the filter can be driven from the mono output available on many decoder amplifiers.

When using these sub-bass filter units, be assured that the main amplifier outputs will not be further loaded, because the bridge input of the filter is a fraction of the speaker impedance; around 20,000 ohms.

The output of the filter is often split into a non-inverting and an inverting output pair, so that it will drive both sides of a stereo amplifier in correct phase for a bridge mono output configuration (page 35). This limits the load impedance to 8 or 16 ohms because the two amplifier outputs are effectively in series, but will give more than double the output of of a single channel in an area where the more available power you have, the better.

It is necessary to connect the sub-bass speaker in phase, so that it adds to the bass output rather than subtracting bass from the other speakers, To ascertain correct phase, try the sub-bass speaker connected first one way, and then the other way round. There's only two ways it can go, and the difference will be obvious.

The low pass filter frequency is set to the point where the bass is extended, not just boosted, the sub just overlapping the main speakers. The sub-bass channel gain similarly provides bass extension rather than putting a lump on the bottom end of the spectrum. The purpose of the sub is not so much to share the task of the other speakers, but to add dimension to the last octave in the spectrum, from 40 to 20 Hz, the region where most speakers have lost their grunt. The sub-bass channel therefore does not boost the bass at a particular point, but extends an otherwise flat response downward to an extended flat response, with enough low distortion output to handle dynamic extremes with effortless power.

Choose a quality sub-bass system by listening for the real subsonics, because trying to force extension of bass range just by turning up the gain will only generate a peak that sounds much like a cardboard box colliding with a stick. The sub-bass doesn't produce an output all the time, and should not be expected to; the program will determine when it should speak, but correct level alignment will blend in the bass extension smoothly and make it an integrated part of the audio spectrum.

BASS DISTORTION AND POWER LOSS

The Inverse Square Law requires more power at the lower frequencies of the bass spectrum. A compact, relatively inefficient speaker combined with small amplifier power of 20 or 30 watts may run into trouble on bass peaks. When an amplifier clips bass waveform, it doesn't just sound a bit ratty, it folds its tent completely because of the greater excursion and longer period of low bass waves on which higher frequencies ride.

This effect can occur on the occasional CD with heavy bass, or the surround channel of a movie soundtrack with strong effects, because surround effects do not always have a normal spectrum, and will show up the system if it can't handle bass peaks.

Another cause of such performance is excessive bass boost or too high a surround level. There is often a tendency to add bass or surround volume to enhance reproduction, but as discussed earlier, once levels are correctly balanced and a normal spectrum has been restored, many problems associated with noise and distortion simply fade away.

Another bass hazard is standing wave generation in small rooms. Room dimensions that are less than the wavelength of the sound frequency propagated by the speaker are subject to alternate zones of high and low level, as reflected waves add and subtract at various positions in the room. Careful bass speaker placement can help.

A separate bass or sub-bass speaker can be positioned independently of full range speakers. One effective way to position it locates the nodes of maximum effect by the following means:

1. Place the speaker at the main listening position and raise it to the height of a seated person's head. It will most likely be perched dangerously on your favorite chair, so bear in mind that soft furnishings slowly settle under the weight, and your expensive bass box will not crash to the floor when your back is turned.

2. Set up a suitable program at moderate volume, and crawl around the room, charting all the locations two feet above the floor where the bass output is strongest. You now have a map of your room with all the places marked where the bass speaker can be placed for full effect, and you have only to position the speaker at the most convenient one to get maximum output at the listening position.

A further occasion for loss is when the speaker is located close to an opening into another area, or near a poorly supported thin panel wall. Some ceiling and wall panels have low mass per unit area, and if they are free to vibrate, they will absorb bass in the same way that the resonant absorber panels of Chapter 9 operate. The remedy is to add mass or extra support to the lightweight panels. As indicated in Chapter 9, boomy conditions in small rooms can also be fixed using the principles of the Acoustic Resonator Absorber.

It will be apparent that your bass speaker setup and placement has been successfully accomplished when your TV viewing has the feel of a cinema, and your CD's seem to have acquired the low frequency presence that you experienced at last week's live performance, which you can not now erase from your memory.

MAINTAINING PROGRAM DYNAMIC RANGE

In Chapter 8, on the subject of volume compression, it was suggested that further processing would probably occur in copying and broadcasting. But nothing that can be done to over-cook the sound compares with the action of **automatic gain control** functions on the average VCR's audio record channel.

The AGC is there for good reason, to bring all programs received to the same level so that recordings will be free of audible system noise and overshoot distortion, but one of its side effects is to spoil the dynamics of program peak levels, holding dramatic music swells and loud effects down to the same level as the rest of the program.

In order to make off-air recordings of soundtracks and music programs with true dynamics, it is necessary to disable the AGC and set average record level about 6 dB lower, so that headroom will be left for the peaks. The only practical way to do this is to purchase a VCR with an AGC override switch: **Audio Recording; AGC/Manual**, together with Left and Right channel gain controls, and a stereo level meter. The VCR with these facilities may be expensive.

The signal to noise ratio of Hi-Fi Stereo audio recordings made on the VCR video scanning heads is adequate for the extended dynamic range gained this way, and the only other recommendation is to lock the position of the gain control sliders with adhesive tape so they won't drift accidentally. A recording made with the AGC active may not reveal a problem until the last scenes of a movie, and to have the volume suddenly pulled down at a time of maximum demand is a frustration that need not be endured. The next VCR you purchase can have the facility to eliminate the problem that spoils the impact of dramatic moments, and is one more step to obtaining the maximum result from the Total Theater you have so carefully set up.

Appendix A

Working with Formulae—
Practical Examples of Common
Calculations

CALCULATING VOLTAGE DROP RESISTANCE

Example:

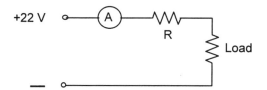

The supply is 22 volts (AC or DC).
The Load specification is 12 volts at half an amp.
A series resistor is required to drop 10 volts.

Covering the wanted quantity on this button reveals the formula:

$\dfrac{V}{I}$ which equals $\dfrac{10}{0.5}$ = **20 Ohms**.

CALCULATING AMPLIFIER OUTPUT POWER

Follow the method given in Chapter 3 and calculate as in the following example:

Measured: 45 volts across an 8 ohm load.

$$\text{Watts} = \frac{V^2}{R} = \frac{45 \times 45}{8} = 253.125 \text{ or } \textbf{253 Watts.}$$

Note that the initial voltage measurement is critical, and requires a meter that reads true RMS value, because the figure will be squared in the calculation. A small inaccuracy will translate into a much larger error in the final result.

For this reason, it pays to round the final figure to the nearest whole number rather than stick to a fraction that may not be exact.

The accuracy of the meter used for the initial measurement is also subject to careful choice, and it has long been the practice to use analog meters with a mirrored scale so that parallax reading error can be avoided.

CALCULATING LED RESISTOR VALUES

Example: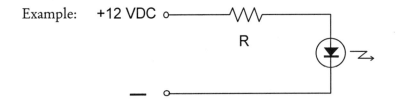

The supply is 12 volts, the LED specification is 1.7 volts at 20 mA, so the series resistor is required to drop 10.3 volts at 20 milliamps.

Covering the wanted quantity on this button reveals the formula: $R = \dfrac{V}{I}$ which equals $\dfrac{10.3 \times 1000}{20} = 515\,\text{Ohm}$

20 milliamps is the maximum current for the LED selected, but to allow a safety margin, use the next highest preferred value resistor: <u>560 Ohms.</u>

CALCULATING AC TRANSFORMER SECONDARY VOLTAGE FOR A GIVEN DC RECTIFIER OUTPUT

Example: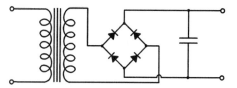

The required DC is 18 volts, the rectifier is full wave, followed by a large value filter capacitor which will sample and hold the peak value of the AC supply, so the transformer secondary RMS voltage will be the DC voltage divided by the RMS to Peak ratio: $\dfrac{18}{1.414}$, which equals: <u>12.7298 VAC</u>.

Bearing in mind that most small transformers are voltage rated under full load, and a 12 volt secondary will output up to 15 volts, the choice of a 12 volt transformer will most likely be suitable in this case.

Appendix B.

Diagram Symbols

To assist in reading and preparing diagrams, the following series of drawings introduces the symbols commonly used in Schematic Diagrams and Block Diagrams. The complexity of each symbol indicates which kind of drawing it is for. Some suit only the fully detailed Schematic, or Circuit Diagram, showing every connection. Others suit both the Schematic and the Block Diagram, which shows equipment modules as symbols or labelled boxes, with main circuit paths drawn as single lines joining them.

The method and limits of Block diagrams, which are discussed in Chapter 7, demonstrate the value of a drawing which is simplified as far as possible so that it gives an overview of a system without going into unnecessary detail. Schematic diagrams of multimeters and networks, by contrast, are shown in Chapters 2 and 5 respectively.

As can be seen from the following examples, there is considerable latitude in the design of symbols, a fact demonstrated by the diversity of styles displayed in drawings from different product manufacturers. They all convey the essence of the systems they portray. Those who work in professional drafting follow a protocol that is not essential in field work. There are national and international standards for drawn symbols, like almost everything else, but technicians should make and use their own symbol designs based on commercial practice, if they are not familiar with the protocols.

So I commend the following ideas to the reader as they were shown to me, and recommend that you design your own symbols provided they are universally recognizable. A simplified but easy to use and very readable drawing system for field technicians is much better than no system at all. Like most areas of endeavor, the people who acquire the best working skills are the ones who work out their own techniques, guided by observing other people in their industry.

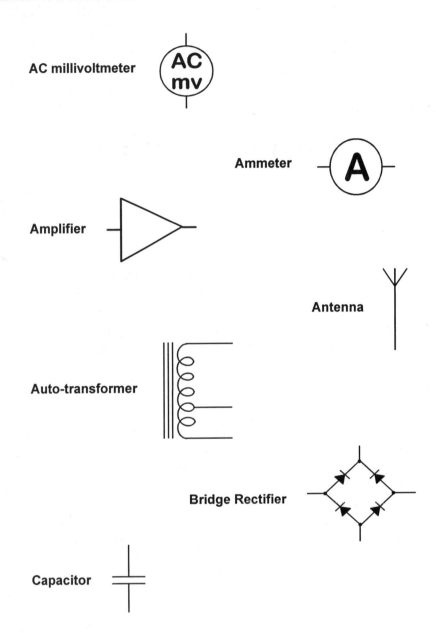

AC millivoltmeter

Ammeter

Amplifier

Antenna

Auto-transformer

Bridge Rectifier

Capacitor

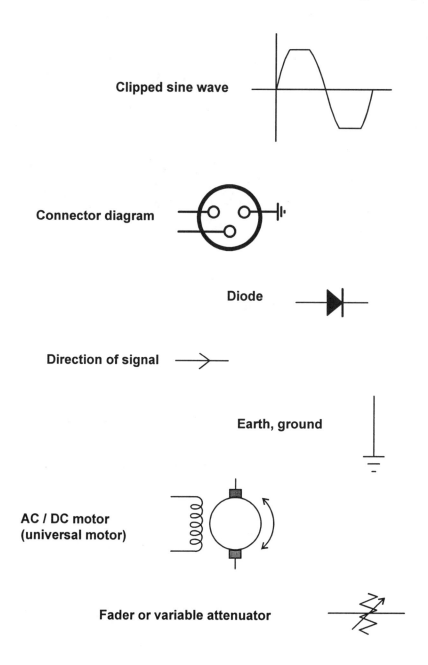

Clipped sine wave

Connector diagram

Diode

Direction of signal

Earth, ground

AC / DC motor
(universal motor)

Fader or variable attenuator

Fuse

Headphones

**Induction motor
(Two Phase shown,
sometimes with a
series capacitor in
one winding for single
phase operation)**

Inductor

Jack

Jacks, normalled pair or insert point

Load or termination

Microphone

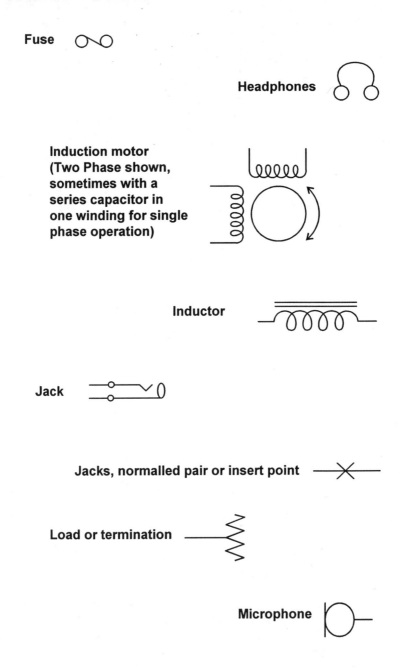

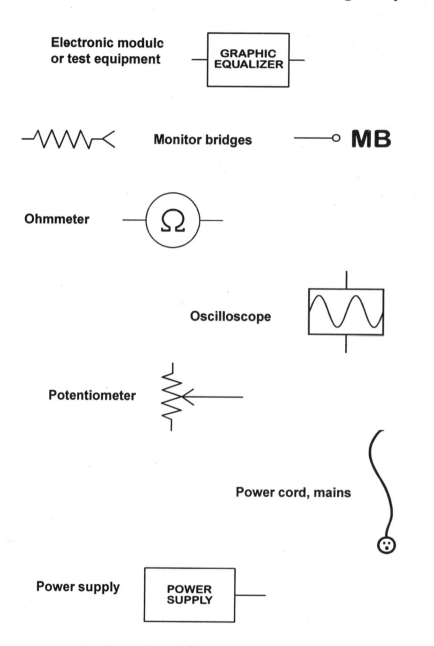

Electronic module or test equipment — GRAPHIC EQUALIZER

Monitor bridges — MB

Ohmmeter — Ω

Oscilloscope

Potentiometer

Power cord, mains

Power supply — POWER SUPPLY

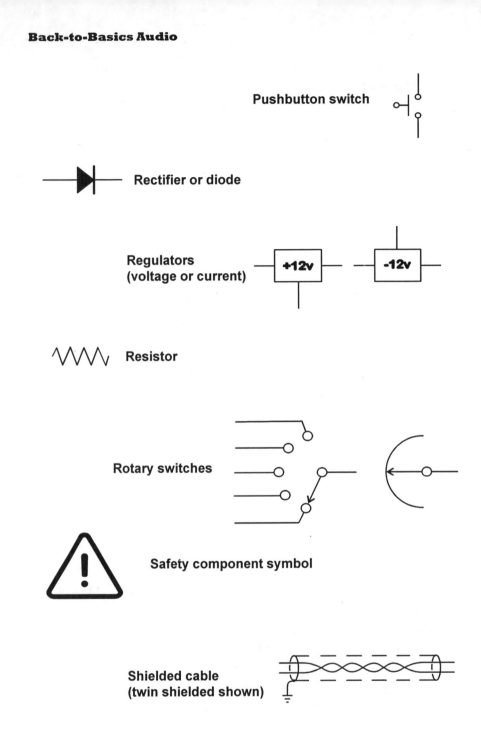

Pushbutton switch

Rectifier or diode

Regulators
(voltage or current)

Resistor

Rotary switches

Safety component symbol

Shielded cable
(twin shielded shown)

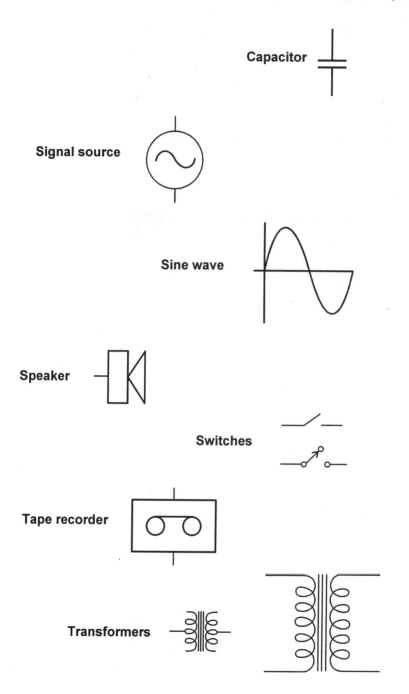

Capacitor

Signal source

Sine wave

Speaker

Switches

Tape recorder

Transformers

Voltmeter, AC or DC

Volume indicator

VU meter (specific Volume Indicator) —|VU|

Appendix C.

A Mechanical Design Exercise

MAKING A HIGH GAIN VIDEO PROJECTION SCREEN

Since the picture is an integral part of a home theater, neither more nor less important than the sound, it is appropriate to observe that if you use a video projector, the screen deserves special consideration, because it will make or break the effectiveness of your presentation. Front projection gives the purest, most real picture of any system, because there is nothing between the image and the viewer by way of shadow masks, lenticular gratings, or diffusing layers. It also places the picture close to the back wall of the room, like a cinema.

Flat screens are very practical, because they roll up and disappear, but a pseudo-spherical high gain screen will at least double the available screen brightness by permitting the use of more reflective screen material. Flat high gain screens, while still practical, inevitably have a hot-spot that follows you as you move around the room, showing a reflection of the light source. Some commercial high gain screens are compound curved so that the hot-spot is enlarged to cover the whole picture area, but they are rigid, scratchable, and difficult to stow away. The semi-curved folding screen shown in Fig. C.1 is made from white pearlized high gain data-screen material on a fabric base. It has withstood handling, cleaning, and continual folding for more than six years.

The ideal high gain screen is a spherical lens that redirects the light from a radial spread to a broad focus that bathes the audience in an oval of illumination, as it is set at an angle that reflects the projector directly into the viewing area. In practice the method is more wasteful than rear projection through a lenticular screen, but the advantage is you can still see the picture quite well over a wide angle, seated, standing, or way off to one side. The semi-curved screen described here works a satisfactory compromise. Screen illumination appears effectively constant right into the corners provided nobody whips out a spot-photometer and looks for errors.

The screen can be concealed in folding furniture using a lever-hinge arrangement to open it at the right angle. The screen shown in the photographs of Fig. C.1 is four feet wide, and is masked to a 60 inch diagonal picture with a ratio of 1.38 to one. The two inch velvet masking is enough to hide top and bottom over-spill without video letterboxing, and while the particular three-gun projector used does not have a scan doubler, the raster is concealed by an effective image-spread phosphor design which doesn't detract from good quality pictures. When the screen opens, a small computer fan pulls air out of the chamber behind it in order to form the compound curve. The screen material is mounted on steel springs backed by airtight vinyl sheet corner pockets so that the corners can move forward as the center moves back. To eliminate wrinkles at the fold point, the screen material requires careful tensioning for about six inches each side of the fold with springs or rubber bands, as the curvature causes it to take up different positions folded and open.

The screen size and position suits smaller rooms; the room in the photographs is six by four meters, and is set up for viewing straight ahead to the center of the picture, a distance of three and a half meters. Although not ideal for daylight use, the screen enables a six hundred lumen projector to produce higher screen brightness than a cinema, giving bright visual effects the power to make your eyes blink. Viewing at night using indirect room lighting dimmed to about 40 candlepower is such that the projector must be set to less than half the available light output.

The quality of the picture is very acceptable, although three-lens projectors require unique alignment. Combined with five channel sound plus sub-bass, it presents the viewer with a potent cinematic experience, the feeling of being part of the scene.

The screen can be built into a variety of furniture styles, like tables or console cabinets, but the lightest in appearance and one of the easiest to hinge is the desk, because it can have legitimate side members rising above the work-top surface to carry one end of the top hinge bar on each side. The handle is on the lower half of the screen frame as that is the only place from which it can be operated by hand, since the top section executes a rise-and-drop motion as it opens. Was it worth going to all this trouble just to fold a curved screen? Yes, absolutely it was worth it.

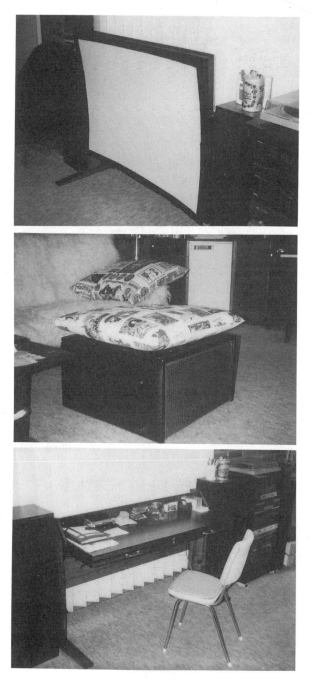

The screen is curved by a computer fan pulling air out of the sealed chamber on which it is mounted.

The projector lives in a box which doubles as a table or a seat.

When the show's over, the screen folds up to become something useful. It has a catcher on the back which automatically tidies away the paperwork.

Fig. C.1

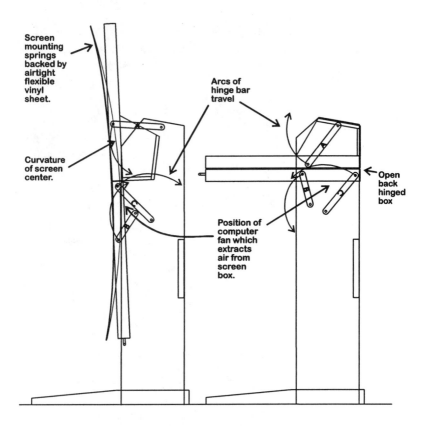

Fig. C.2 Construction of the curved screen, shown open,
and closed to make a desk.

CONSEQUENTIAL DESIGN STEPS

A large part of mechanical design is involved with things that have
interdependent movements; like levers, and double action hinges. Many
mechanisms operate through angles, and to determine their action, arcs
are drawn to show the start and end positions.

The Block Diagram technique described in Chapter 14 creates a path-
way to essential information which would be difficult to gather in a more
direct way. Designs like the folding screen are the same; unless arcs are
drawn, the final design will be much more difficult to achieve. When in
doubt, draw a diagram showing the travel arcs.

This design method tells us something else. If it was intended to
motorize the screen, the best place for a chain-drive sprocket would be at
the bottom end of hinge-lever 'C', shown in Fig. C.2, since it is the

member that moves through the greatest angle, and is clear of other moving parts at all times. It is also in a place that is conveniently hidden under the desk, as linking the left and right levers at either side of the screen with a torsion bar would ensure smooth, jam-free operation.

Front opening racks can also use double-lever hinges, so it's a technique worth experimenting with. Before committing this type of design to metal and wood, full size, it is a very good idea to prove it works by making a flat cardboard model of the hinge on a piece of wood, with pins at the pivot points. The drawing in Fig. C.2 was traced off a photocopy of the fifth-scale model of the mechanism.

Appendix D.

Estimating Power to Speakers

It's one of those hard facts of life that once a speaker voice coil is burned out, it cannot be fixed and the whole driver usually has to be replaced. Also, the program from that particular channel or frequency band comes to a halt, and action has to be taken to restore it to keep the show going.

Avoiding this sort of trouble is the subject of much advice and speculation, whether you are engaged in using, selling, hiring out, or servicing the equipment in question.

Although a line fuse may not save a speaker from "slow burn" overload or hazardous waveforms, it will often save it from sudden accidental overload, as can happen when a microphone channel is inadvertently left open. Speaker manufacturers will advise on suitable ratings, but 4 amps for every 250 watts into 8 ohms is usual. An automotive in-line fuseholder size 3AG is the quickest way to fit the fuse at a positive speaker terminal or the amplifier output connector.

But there's a very simple method that will prevent a high percentage of such disasters, and that is to measure the RMS signal voltage applied to the speakers.

As discussed early in Chapter 6, there are other situations that result in speaker damage, like choice of too small an amplifier resulting in a clipped signal, and amplifier faults that may pass direct current through the speaker, but one frequent problem I've encountered is that the owner or hirer of a speaker system often has no real idea how much power is going to the speakers.

The information in the following two pages is intended to take the pain out of buying or renting speakers. It applies equally to domestic and professional users, because the causes and results are much the same.

HOW MUCH POWER IS GOING TO THE SPEAKERS ?

Speaker burnout causes trouble and expense. It can be due to:

1. **DC on the speaker line.** This can be the result of an amplifier fault even if it still functions. It sometimes manifests itself by burning only one end of the voice coil: the part that was pulled clear of the magnet. It overheats because that end of the voice coil then has no heat sink to the magnet yoke, and the coil also has lower impedance.
2. **Clipping.** Over-driving a line or power amplifier generates square waves that are not always audible from bass speakers. For this reason, using under-powered amplifiers is often a greater hazard than too much undistorted output. A clipped signal produces more effective energy than the amplifier's rated output because it compresses the waveform, increasing its duty cycle, or on/off ratio. This is also damaging because it contains high frequencies above normal levels that the speaker cannot follow, resulting in overheating and mechanical stress.
3. **Over-driving.** Excessive output from a large amplifier will damage speakers. Use of an output sensing limiter will help to solve the problem by preventing over-driving and/or clipping.

FIELD ASSESSMENT OF POWER TO A SPEAKER

If a speaker line sensing limiter is not available, power to the speaker can be calculated from the RMS voltage measurement of the speaker line and the nominal impedance of the speaker. Ordinary multimeters may have poor high frequency response, but will be reasonably accurate at low frequencies.

Field assessment is simple. The AC voltage measurement of the program signal should be taken at the peak swing of the meter. Using the power formula: *Watts equals Volts Squared divided by Impedance*, the following table will approximate the power to the speaker:

RMS volts	Watts in 8 Ohms	Watts in 4 Ohms
20	50	100
30	112	225
40	200	400
50	312	625
60	450	900
70	612	1225

The indicated power is a useful guide provided the signal has normal aspects of compression and frequency response. Bass boost or volume compression will reduce the power handling of the speaker by the amount of processing in decibels.

Refer to Chapter 3 for amplifier testing and output power measurement, using the same power formula.

Glossary — Index

3 dB loss point, defines the cut-off frequency of a filter, 170, 171 (Fig. 8.10)
70mm film sound. Six track magnetic stripe recorded with noise reduction. 125
70 volt line, constant volt speaker lines. 34
100 volt line, constant volt speaker lines. 34
600 ohm line, standard impedance audio line. 46, 59, 62, 64
630 Hz, center of the audio spectrum, a program balancing tool. 54, 109, 169

A

A chain, a sound system up to line level before the main gain control. 128
A/B monitor, direct and simultaneous tape replay monitor, comparison of two signal sources. 150 (Fig. 7.10), 151
Academy roll-off, original cinema acoustic response standard for monitoring and reproducing monaural soundtracks. 124 (Fig. 6.17)
Absorber resonator panel, for low frequency acoustic control. 111, 181-184
Absorption, acoustic, the inverse of reflectivity. 179
Absorption spectrum, frequency response of a room's non-reflective properties. 179
AC millivoltmeter, an amplified voltmeter. 18, 62, 91
Acid-gel battery, non-spill low maintenance primary (rechargeable) lead-acid battery. 232
Acoustic absorbers, commercial panels, for reverberation control. 185, 186
Acoustic brighteners, mid and high frequency acoustic reflectors. 190
Acoustic climate, reverberant and reflective properties of a room. 179, 180
Acoustic flat absorption response, featuring combined treatments. 182, Figs. 9.3 and 9.4
Acoustic mid and high frequency absorption using diffuser panels. 182, Fig. 9.4
Acoustics, study of sound in a reflective environment. 179, 180, (Fig. 9.1)

Adding noise or distortion figures. 59
AGC, automatic gain control. 307
AF, audio frequency. 53
Airflow, forced fan ventilation. 186-188, 195, 196
Air suspension, sealed speaker box design. 113
All-pass filter, neutral network that controls phase and delay. 88
Alternator, alternating current generator. 194 to 196
AM, amplitude modulation. 55
Ambient noise, background noise in a listening or recording environment. 167
Ammeter, amp meter, measures amps (amperes). 19, 20
Amp, ampere, the unit of electric current flow. 3
Amplifier instability, caused by externally linked input and output neutrals. 146
Amplifier, converts signal input voltage to output voltage and current that drives lines and speakers. 29
Amplifier, voltage amplifier, recovery amplifier, converts signal input voltage to a higher output voltage at the same impedance. 47, 94
Analog (sound), continuously variable signal format, as opposed to Digital, which uses a numeric code to address waveform levels. 175
Analysis by diagram, drawing block-diagrams, shortcut to understanding complex systems. 137-144
Angstrom, unit of length used to describe wavelengths of light, a ten millionth of a millimeter. 56
Arc lamp, plasma discharge, incandescent conducting ionized gas. 2, 13
Assemble-edit, drop-edit, electronic tape edit in magnetic recording. 148, 151, 173
Atmosphere recording, ambient background sound recording to inter-cut or mix with with edited tapes. 167
Atmospheric pressure. 196
Attack and release times, audio envelope of

Z

Zener diode, regulator component conducts
above zener voltage. 228
Zero-crossing, AC switching timed to occur at
zero volts portion of waveform. 204
Zero level, zero VU, test tone, relationship with
line level. 152
Zero setting of a fader, compared to full gain,
usually allows for headroom of 6 to 10 dB
above unity gain setting of mixer, fader is said
to "hold" 6 or 10 dB in addition to its insertion
loss of about 6 dB. 61 (Fig. 4.5), 160